GREAT LENGTHS

A Celebration of the Surveyors of Ontario

CHARLES WILKINS

Map of
UPPER & LOWER CAN
with the
DISTRICTS & TOWNSHI
into which the Provinces are divided.
Diagram of a Township, shewing the Crown & Clergy Reserves.
UPPER CANADA.
Lake Huron
Lake Ontario
Lake Erie
L. St. Clair
Georgian Bay
Indian Territory
French R.
Ottawa or Grand R.
L. Nipissing
Temiscaming
L. Simcoe
Severn R.
Newcastle
Midland
Home
Gore
Niagara
Goderich
Dundas
York
Lockport
Rochester
Buffalo
Erie
Long Point
Detroit
Amherstburg
Chatham
Harbour
Oswego Canal
Erie Canal
State of Michigan
State of Pensylvania
State of Ohio
State of

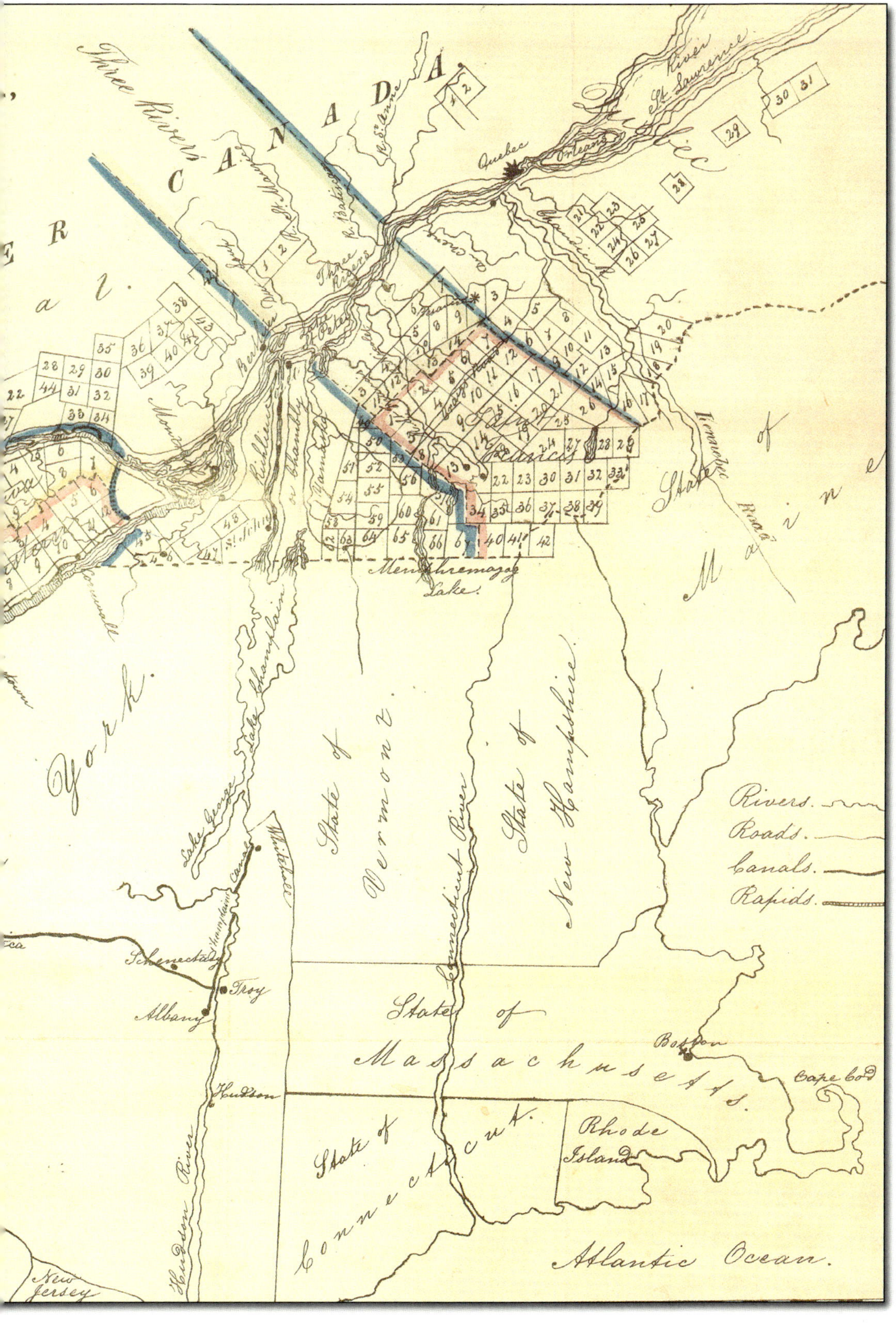

The Province of Upper Canada existed for 50 years, from 1791 to 1841, and generally comprised what is now southern Ontario. During its relatively brief time on the map, it evolved from undifferentiated forest, broken by occasional clearings, to a kind of crazy quilt of districts, townships, municipalities, farms, urban lots, and of course streets and roads and ever-shrinking "Indian Lands." The land surveyors of the day were pivotal to this relentless transformation and to the subsequent building of the much larger province of Ontario. Note the checkerboard pattern at left, showing the ratio of township lots reserved for the clergy (black rectangles) to those reserved for farmers or other citizens (white rectangles).

www.aols.org

Softcover ISBN 978-1-894801-32-4

Hardcover ISBN 978-1-894801-33-1

Contributors

AOLS Book Task Force: Vicky Culbert, Blain Martin, Maureen Mountjoy, Ron Stewart, Andrea Tieman

Copy Editor and Fact Checker: Dawn Promislow

Surveyors Register Checking: Bill Buck, Bob Mountjoy

Interview Transcriber: Narineh Kassabian

Proofreaders: Doug Aron, Bill Buck, Maureen Mountjoy, Ron Stewart, Andrea Tieman

Printed in the United States of America by LSC Communications, Crawfordsville, Indiana

Design and Production Management: Dan Diamond and Associates, Inc.
Toronto, Ontario, Canada

Photo Credits: see page 215

Table of Contents

An intrepid crewman enjoys the view from the top of a lookout ladder during a maintenance assessment of the Canada-U.S. Border, at Rainy River, Ontario, circa 1948.

credit: International Boundary Commission

Profiles & Waypoints

For the members of the
Association of Ontario Land Surveyors,
past and present

Association of Ontario Land Surveyors

Hello and ...

... Welcome to *Great Lengths*, the story of the Association of Ontario Land Surveyors, published to mark our 125th anniversary in 2017. This book does many things well, but most importantly, it tells the story of who we are – as a profession and as individuals.

In the course of the book's development, our writer, Charlie Wilkins, interviewed more than 150 surveyors, collecting their stories, photos and artifacts, all the while building a perspective on the history and common interests that bind them. Charlie's research extended to libraries, public archives, government offices and art collections – and of course to the corners of the Association's own growing archive. His introductory essay, "A Year Among the Surveyors," connects everything from the exploits of the earliest surveyors in Upper Canada to the Erindale Survey Science program of the 1970s to the latest in digital technology and the passions and adventures of our youngest surveyors.

Modern surveying is a life of outdoor stimulation, advancing science, and the need to understand legal principles, social forces and communications. I am delighted that *Great Lengths* will serve to introduce young people to the exciting frontiers of our profession. The generosity of our sponsors has enabled us to place a copy of the book in university, college and high school libraries throughout the province.

A project of this magnitude requires the support of many participants. I want to thank the AOLS Council for its approval, the AOLS book task force and Archival Committee, the sponsors that made this ambitious book financially possible and, of course, the individual AOLS members who contributed in such a variety of ways. Beyond all of that, I want to express a debt of gratitude to our author, Charlie Wilkins, for his vision of what the book could be, his dedication to that vision, and his skilled writing along the way – all very much appreciated.

Now, it's time for us to enjoy *Great Lengths*.

Sincerely,

Blain Martin
Executive Director
Association of Ontario Land Surveyors

1043 McNicoll Avenue ♦ Toronto, Ontario, Canada ♦ M1W 3W6
Tel: (416) 491-9020 ♦ Fax: (416) 491-2576 ♦ Toll-free: 1-800-268-0718 ♦ email: admin@aols.org

A Year Among The Surveyors

Vaitheki "Vicky" Kumar, OLS #1986, at work in the Toronto area for her employer, J.D. Barnes Limited.

Gordon Gracie, shortly after his arrival at Erindale College in 1971.

ON THE MORNING OF TUESDAY, MARCH 30, 1971, A GIFTED, CANADIAN-BORN ENGINEER NAMED GORDON GRACIE received an unexpected letter from John Tuzo Wilson, one of the world's most accomplished geophysicists. A decade earlier, Wilson had established a radical new course for global geological science with his insights into the movement and mysteries of the earth's tectonic plates. However, his letter to Gracie came not from any geophysical lab or research site but from the desk Wilson now occupied as the principal of Erindale College, a newly established outpost of the University of Toronto.

Gracie, a former professor of engineering at the University of Illinois, had for the past six years worked for the Raytheon Corporation, an aerial photography firm in Alexandria, Virginia, a suburb of Washington, D.C. Among his responsibilities with Raytheon was teaching the uses of aerial photography to CIA recruits, at a secret location in Washington, where his comings, goings and academic endeavours were closely monitored by the FBI.

Such monitoring may in a small way have factored into Gracie's feelings that at some point in the foreseeable future he would like to move his career in a different direction.

During an interview in May 2015, at his home in Burlington, Ontario, Gracie recalled reading Wilson's letter on that morning 44 years earlier and the excitement he felt at the possibilities suggested in its paragraphs. It said:

> *Dear Dr. Gracie;*
> *I understand from Professor O.J. Marshall that you may be moving your location and might be interested in returning to academic life.*
> *We have not yet advertised any position, but we are contemplating the possibility of teaching surveying at Erindale College as part of a broad science program.*
> *If you are interested in a meeting, we should be glad to pay your expenses if you would care to visit us. ...*

The letter went on to describe the character of the college, its location, and its relationship to the University of Toronto – and then to say:

> *If you find you can come to Toronto we can meet you at the airport and arrange accommodation. I imagine a one-day visit would be sufficient and this could take place on a Saturday, although another day would be better.*
> *I look forward to the possibility of meeting you.*
>
> *Yours sincerely,*
> *J.T. Wilson,*
> *Principal*

Beneath his signature, in a tiny, hand-inked scrawl, the renowned geophysicist

The new Erindale campus, seen at left, during the late 1970s. At right, Erindale students (left to right) Ann-Marie Mailhot, Norm Beck and Kathy Sam practice their surveying technique at Ox Narrows, Haliburton, 1978.

urged Gracie to phone him (*here is the number; do it now,* sort of thing).

So it was that three days later, at 7 a.m., at Washington National Airport, Gracie boarded Alleghany Flight 691 to Malton Airport in Toronto. In the bottom right corner of Wilson's letter, which Gracie carried in his brief case, Gracie had penned a reminder to himself that upon his arrival in Toronto at 10 a.m. he should seek out "a Blue Station Wagon in the arrivals area" beyond the terminal.

The station wagon, as it turned out, was Wilson's private vehicle, and when Wilson had received his passenger and welcomed him heartily, the pair set off down Highway 27 for Erindale. Just four years old, the college was located in what was then a rather remote tract of maple and oak forest in the new municipality of Mississauga, about 30 kilometres west of the main U of T campus.

There, during an afternoon of meetings and queries, as well as a meandering inspection of the facilities, Gordon Gracie, 41 years old and in his prime, glimpsed a vision of his future, and possibly the future of surveying in Ontario – and privately committed to that vision.

It was a commitment that three months hence would see him installed as director of the new Erindale Survey Science program – a program that for Ontario surveyors today has about it an aura almost of legend, not to mention of the lost Atlantis (a likeness that we shall visit in a subsequent chapter of this book).

Meanwhile, Gracie's visit to Toronto marked not the end but the mere midpoint of nearly three years of frenetic planning by, among others, the Association of Ontario Land Surveyors, along with a variety of commited academics and surveyors, and a gaggle of agents from the provincial Ministry of Education.

The aim of all this planning was to create a four-year university program devoted entirely to the diverse interests and requirements of surveyors – a program the likes of which had never existed in Ontario. Prior to this, surveyors had absorbed their education piecemeal, sometimes with the backing of, say, the Department of Highways or Ontario Hydro or the Department of Lands and Forests; or by taking a university engineering program with surveying options – all the while doing field work, articling, attending survey camps and, eventually, writing final examinations set by the AOLS.

According to Kenton Campbell, a graduate of the Erindale program, Gracie's commitment and the planning that preceded it ushered in what Campbell calls "the golden era of modern surveying in Ontario."

Given the subtle presence that surveying enjoys in the life of the culture

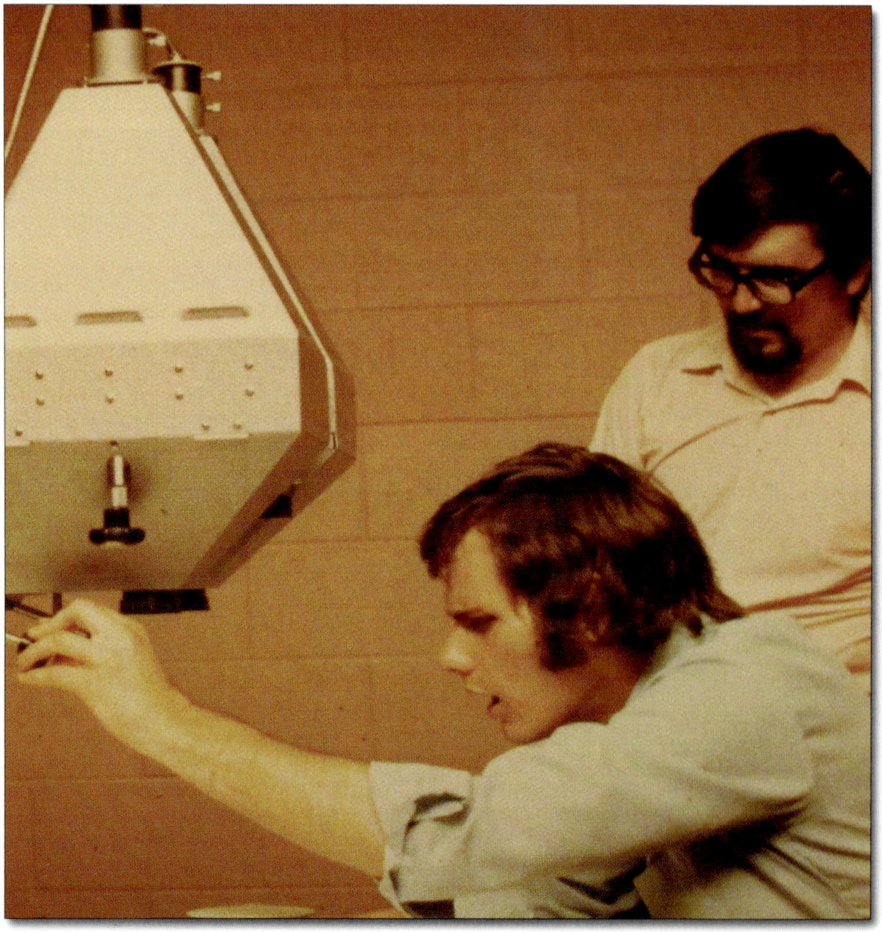

Left: Doug Aron (foreground) and Howard Turner hone their photogrammetry skills at Erindale, circa 1978. Brian Maloney, seen above in 1979, was the first Erindale grad to become Surveyor General of Ontario, a post he held from 1998 to 2010.

("ghosting among the touchstones of the observable world," as Henry David Thoreau characterized the profession), some might be surprised to hear that surveying has even had a "golden era," much less a modern one. However, as the pages that follow will make clear, it has had not one but several, all of them deserving of the "golden" designation.

During the twenty-five years of its existence, the Erindale program would turn out some 450 graduates – a corps of surveyors that, even today, 18 years after the last of those grads fitted mortar boards to their fertile skulls, makes up a majority of the province's 650 active and retired land surveyors. Erindale grads have been among the eminences and interlocutors of the Association of Ontario Land Surveyors; have been Surveyors General for the province; have become professors and examiners, and Association and corporate executives. They have travelled the world in support of vital infrastructure projects in developing countries and have done equivalent service on projects here in Canada. Collectively, they have constituted a walloping majority of the practicing surveyors of Ontario during the most advanced period of surveying in the province's history.

More pertinent to the project at hand, the "Gracie Cohort" has been a kind of bridge between the methods and mentality that eased surveying out of the war years into the flourishing mid-century and, further, into a world whose practices and technology had, by 1998, when the program went dark, altered the face of surveying.

All of which places Erindale squarely in the transit sights of the book you hold in your hands – a book that is itself a kind of bridge linking the past and present and future.

Walk to the centre of that bridge, swing your transit, and you are looking

Eighteenth-century mapmaker Peter Pond, seen here as depicted by the artist Illona Campbell in 1984. The scene (described by historian, Philip Vail) is from the first meeting of Pond and his protege, the explorer, Alexander Mackenzie: "Pond stalked into the hall, a pack of dogs at his heels. The grey-haired giant had not shaved in weeks, his buckskins were stained, and he was badly in need of a bath. But his natural dignity was overwhelming. He ate a large venison steak, a platter of bear-bacon, and a moose liver – and insisted his dogs be given fresh meat, too."

at a history so charged with drama and escapade, with exceptional human ingenuity, that Hollywood could scarcely have invented it.

Consider for a moment the cast of characters who brought that history to life – men who with little regard for their safety (or sometimes, it seemed, their sanity) trekked the marshes and forests and frozen rivers, laying out the townships, farms and roads, the train tracks and bridges and navigation dams, on which modern Ontario is founded. The earliest of those resourceful irregulars were dreamers, adventurers, survivalists, many of them as tough as the 66-foot chains that they hauled around to measure lots and road allowances and prospecting patents.

Many of them, it should be noted, were also functionaries of a prolonged colonial incursion, largely British, that was by no means sympathetic to the First Nations or their interests (about which we shall have more to say).

They were men such as Peter Pond – mapmaker, fur trader, murderer – a mysterious and tough-minded speculator who did much of his plotting and mapping in the Athabasca region of the far north, but also plied the rivers and forests in what is now northwestern Ontario.

And Samuel Holland, Upper Canada's original Surveyor General, a man so steadfast in his administrative responsibilities that he worked daily for six penniless years as he waited for the government to issue him his first pay cheque.

And Augustus Jones, who surveyed more than a quarter of Ontario's townships prior to 1800, and was considered the elite surveyor of his day, despite being denounced as an "adulterous poltroon" by the lieutenant governor of Upper Canada, Sir Francis Bond Head.

There was David Thompson, who in his surveying and map-making trudged more than 90,000 kilometres through the North American wilderness, and mapped nearly 5 million square kilometres of it. Back in Ontario during the early 1830s – going blind, going broke, going quietly snaky in his superannuation – the greatest surveyor who ever lived resurrected his career just long enough to do an arduous survey of Lake Muskoka, which brought barely enough income to prevent him and his wife Charlotte from falling into destitution.

Alexander Niven, surveyor of the famous Niven's Meridian, was another. As was Sandford Fleming, who surveyed what is now the CNR line across northern Ontario and subsequently divided the planet into time zones, giving the world the enduring and ever-functional Standard Time.

David Gibson was yet another – an intrepid practitioner, whose association with William Lyon Mackenzie and the Rebellion of 1837 got him chased from Toronto, where his home and belongings were ingloriously torched behind him.

Exactly a decade later, Publius Virgilius Elmore, a surveyor as obscure as Gibson was notorious, laid out the first Colonization Road into the Ottawa-Huron Tract, slogging at times through waist-deep swamp, battling what he referred to as "hecatombs of murderous black flies." The tract, a vast swath of mixed forest and lowland extending from Georgian Bay eastward to the Ottawa River, is worth mentioning

While articling as a surveyor in 1837, Henry Strange (top left), a colonel in the Canadian military, seen here in his silk scarf and lambskin hat, took up arms against another surveyor, David Gibson (far left) whose offence against Strange and the Canadian army was that he was a participant with William Lyon Mackenzie in the Rebellion of 1837. It is the only recorded case of land surveyors in Ontario going into armed conflict against one another.

because, like so much of the territory settled in Upper Canada during the early and middle 1800s, it was appropriated by the colonial government from the First Nations, in this case from the Chippewa, for a criminally small outlay of perhaps five cents an acre.

Quite apart from swamps and black flies (and the churlish pen of a writer nearly two centuries after the fact), Mr. Elmore bore up under what were surely the most pompous and eccentric Christian names in the history of Ontario surveying – and, moreover, under the daftest hairdo, a kind of geodetic comb-over, swept forward with logarithmic precision from above each temple onto the curvature of his otherwise featureless pate.

In the days just after I began reading about these extraordinary early surveyors, I found myself in conversation with the Stratford surveyor Peter Moreton about the conspicuous and seemingly endless idiosyncrasies of bush lords such as Pond and Jones and Elmore. At one point in that conversation, Peter observed that decades ago, when the AOLS was assigning numbers to the earliest Ontario Land Surveyors, it was "fitting that those doing the assigning gave Number One to a man named Strange" – specifically, Colonel Henry Strange, who began surveying in 1838.

If Strange's name was apt code for the quirks and foibles of certain early surveyors, the man himself, it needs saying, was a model representative of the solidest traits of those surveyors, as well as a hardy factotum on which to build the numbering system. Born in the West Indies, the Association 's number one man was impeccably educated in England, spoke seven languages, played a dozen musical instruments, was an officer in the Canadian army, and worked as both a magistrate and a surveyor, in Guelph, where he occupied offices in the Priory Building, now the site of a parking lot at Macdonnell and Woolwich streets. Mr. Strange was still surveying at the age of 90, in 1905, and was often seen on the streets of Guelph, carrying his theodolite and, in winter at least, wearing a jaunty Russian hat and silk scarf.

In honour of his numerical primacy, a large and vivid portrait of Mr. Strange,

The given names of the 19th-century Belleville surveyor, Publius Virgilius Elmore, honoured the Roman poet Publius Virgilius Maro, commonly known as Virgil (below). One of the poet's great works, The Georgics, *a series of long poems about the soil and its bounty, found an echo in Elmore's own love of the land and soil, and in his extensive subdivision of the townships around Belleville, and beyond, into hundred-acre farms.*

painted in the style of the Dutch Masters, hangs in the stairwell of the AOLS offices on McNicoll Avenue in east Toronto.

The suggestion that a surveyor's number was occasionally assigned with mischievous intent is born out by the allocation of the number 666, the apocalyptic symbol of the Book of Revelation, to a surveyor named Arthur Hector Death, of Brampton, commissioned in 1948 and active early in his career in the firm Kitto and Death.

By the mid-20th century, the profession was still sufficiently untrammelled to attract and welcome talented outliers such as Neil Simpson, Talson Rody and the late Harvey Hall, a man reputed to be among the most impulsive and eccentric, as well as one of the more brilliant surveyors in AOLS history. The story is told of how in 1936, at the age of 22, on the main street of his hometown of Belleville, Hall happened upon a recruitment booth for the Republican forces in the Spanish Civil War (the Republicans had located in Belleville with the hope of attracting pilots from the air force base at nearby Trenton). Hall is said to have signed up on the spot and, without returning home even to inform his family, to have boarded a train for Montreal, and from there a ship bound for England, then Spain, where, during the course of the war, he took his lumps fighting General Franco and befriended a plain-spoken American journalist named Ernest Miller Hemingway. Despite suffering severe injuries, he joined the Corps of Royal Canadian Engineers in 1939 and throughout World War II fought with the Second Canadian Field Survey Company, gaining a reputation for daring, at times all but manic, enthusiasm for confronting the Germans.

Kenton Campbell tells the story of how in the 1950s, Hall (by this time Ontario Land Surveyor #648) took a flight in a one-engine float plane to a work site in the northern wilds. As the plane did a banked loop over the lake on which it was about to land, Hall, who detested restrictions, both social and corporeal, and who had resisted fastening his seatbelt or properly latching the passenger door of the plane, tumbled out, bouncing off the

AOLS swashbuckler and renegade, Harvey Hall (far left) fought in the Spanish Civil War, where he became buddies with Ernest Hemingway, who is seen above, in 1936, with the Spanish Republican forces. During the 1950s, Hall enhanced his considerable legend as a surveyor with his brilliant transformation of the early town plan of Kitchener, which until 1916 had been called Berlin.

float and pinwheeling a hundred feet or more, without injury, into the lake below. "I *believe*," said Campbell, straining for diplomacy, "that he *may* have been in a somewhat relaxed state at the time. But I'll leave that to the reader's interpretation."

As is often the case with free spirits, Hall possessed a fertile (and in his case highly functional) imagination. During the early 1950s, as chief surveyor for the city of Kitchener, he took it upon himself to overhaul the city's hopelessly haphazard and disjointed survey plan. The existing mess was said to be a legacy of "the German method," by which roads (originally wagon trails) had evolved along convenient or "natural" routes, rather than on an organized grid of the sort apparently favoured by the British. During two years of artful surveying and drafting, Hall reinvented the plan in accordance with the Geodetic Survey of Canada, imposing a grid upon it, tying in every irregular intersection and crooked street, as well as squaring and renumbering block after block of city lots. The result was an exemplary model of consistency that, seventy years later, is still used for infrastructure and cadastral work in the city.

As for Simpson and Rody, I had the pleasure of meeting both men at the Annual General Meeting of the AOLS, at Deerhurst Resort, near Huntsville, Ontario, in February 2015. We shall meet the amiable and tenacious Mr. Rody shortly. Meanwhile, I wish to note that from the moment I heard about Neil Simpson I was keen to make his acquaintance. I had been told by another northern surveyor, Anne Cole, that one of Simpson's abiding ambitions was to paddle every river in Algonquin Park from its headwaters to its mouth, in a birchbark canoe of his own making. It seemed a reasonable ambition for a man who had done just about everything else a surveyor could do to get himself into and out of trouble in the bush over the years. What nobody bothered to tell me was that he was 95 years old.

On the morning we met, in the capacious dining room at Deerhurst, I mentioned to him immediately that I had heard about his Algonquin dream, which I suspected (and he confirmed) would take several years to fulfill; after all, five major watersheds have their sources in the park. As I spoke, he adjusted his hearing aid, looked at me with his pale blue eyes and, after a longish pause, said (as one might say such a thing to a keen grade 3 student), "Uh-huh. You bet. That's what I intend to do."

Toward the end of our 40-minute conversation, Neil told me a story that explains in part how a man with gumption and a knowledge of surveying could, during the mid-20th century make a magnificent wad of money in short order if he happened to be in the right place at the right time (which Neil often was, in that he was a bush pilot with his own plane and could move fast and far when he had to). The story in brief was that, in 1952, when uranium claims were being staked at the rate of several dozen a day around Elliot Lake, Neil showed up in the area to survey new claims (it behooves his legend to report that, at the time, he bore a formidable physical likeness to the professional wrestler Gene Kiniski). For the stakeholders, a survey was a legal necessity if they were to open their claims to exploration. "I could survey maybe one claim a day on average," Neil said, "and, depending on the toughness of the terrain, would charge anywhere up to a thousand dollars to do it" – a rate that in today's dollars would swell to perhaps fifteen thousand daily.

Neil's most memorable job in the territory came when he heard about a New York City speculator who controlled several dozen claims. "I've always been very forthright in marketing my skills," Neil said, "So I risked ten bucks phoning this guy's office in Manhattan, and told him I was in the area and would be happy to survey his claims. Pretty much all he said was, 'Well, thanks very much, Neil! What's the charge?' I said 'Send me a cheque for fifty thousand dollars immediately, and I'll get on it.' The cheque arrived Special Delivery two days later."

At this point in our developing tale, we are going to bid a provisional farewell to the past, take a step or two further onto the bridge, and swing the transit in a new direction. And take a peek.

As in a speculative novel where time has slipped its restraints, we are no longer looking at the folks and forces that got us to where we are but at the young men and women who will lead surveying into the decades ahead, perhaps into the next dimension. They are a group that, like their forebears, will live out their own compelling stories, invent their own rich lore and ideals; will do it with LiDAR and GPS and Unmanned Aerial Vehicles; with castles of data and ribbons of coding – with laser-inspired images so detailed and alluring as to constitute castles unto themselves. They will (if the futurologists are correct) work beneath oceans and quite possibly in outer space; will work from vehicles and conveyances that technology suggests will drive and fly themselves, perhaps even without the roads, rails and airports that the surveyors of the 20th century spent their lives developing and maintaining. They will drive the profession from both ends of the gender spec-

Geiger counters, such as the Model 107C Professional seen above, were a common sight in Elliot Lake during the early 1950s, when the flying surveyor, Neil Simpson, descended on the territory with his transit and steel tape.

Talson Rody • The Sweet-smelling, Sweet-sounding Hereafter

"**Once during late November in I guess the mid-1960s,** a guy named Frank Dunn and I were up northwest of Timmins, doing a boundary resurvey in heavy bush. We were travelling with a muskeg tractor, pulling what we called a sloop, a big sled with steel tubing for runners. We were on an old logging road, and came to an area where the road, which was corduroy at this point, was imbedded in maybe six inches of ice that had formed on the marshy part of a beaver pond. We were afraid the ice might not support the tractor and sloop if we happened to slide sideways.

"After thinking about it for a while, we edged the tractor forward, and sure enough the sloop dipped to one side, then off the road, and we had to get out on the frozen pond and try to lever it back on.

"The two of us were well out on the ice lifting straight up with a heavy spruce pole we'd cut, and suddenly the ice gave way under us and we were propelled like rockets down into the water and muck.

"What made it worse was that we were directly behind the beaver dam, and in winter there's no place for the beaver waste to go beneath the ice; it's all just down there, a big slew of sewage. Oh, it was awful; we were completely soaked with this disgusting feces. But we had no other clothes to change into, so we just had to keep going.

"When we got to the survey site it was nearly dark. The only shelter was a rotting tent with a plywood floor from an old mining exploration – and a light stove but no wood cut. We had our own tents, but we were so exhausted and cold at this point we just took off all our clothes and crawled into our down sleeping bags.

"I don't have a particularly strong personality, but I think I have strong character. Whenever there was something unpleasant to do out in the bush, I'd try to do it myself, because I didn't want to have one of my workers refuse to do it.

"By dawn, I'd had a fairly comfortable night naked in my sleeping bag. But with all the beaver waste saturating my clothes, they were frozen solid; you could have cut them into slices with a chain saw. The only thing I was able to pull on was a big pair of frozen boots. So, out I went, with an axe, stark naked in the cold. I was out there 20 minutes or so – pretty much every part of me was blue. But I did get some wood cut and we got a fire going – got our clothes thawed and dried. But not cleaned of course; we had to do the entire survey in these wretched, smelly garments – definitely not what you'd call a sweet memory of my career."

"Another time, in maybe 1964, I had a crew out in the French River area, doing surveys for a hydro line. We were staying in a motel south of Sudbury, travelling to the site by car and then downriver toward Georgian Bay in a quite heavy metal powerboat.

"One day by quitting time we'd moved quite a ways along the riverbank from where we'd started that morning. I sent one of the fellas upstream to fetch the boat. But somehow as he was accelerating out of the mooring spot the engine flipped off the transom; he hadn't checked to see if it was bolted on properly. We didn't have a hope of finding it in the deep water, and ended up having to paddle that heavy boat all the way back upriver to the highway, maybe 15 miles. By the time we reached the motel, we were half dead. I don't think we even ate – just fell into bed.

"The next morning it was pissing down rain – I could see and hear it being blown against my window. One of the guys stuck his head into my room to ask what we should do. I said, "Go back to bed – nothing doing in the rain.

"Only later did I realize that one of my crewmen, Cecil Steepe, was the rainmaker – he was out there with a garden hose, flicking the spray off my window, while the other guy solicited my opinion. I slept like a baby for another three or four hours. As far as I remember, it was the only time in my career that I was quite happy to be made a fool of by my crew."

During the next 36 years, under Talson, the firm conducted dozens of regional resurveys, scores of mapping projects for northern municipalities, and plotted hundreds of miles of line for Ontario Hydro, jobs that often called on Talson's skills as a pilot.

Beyond his talents as a surveyor, Talson is renowned for the fairness and appreciation he invariably showed his crews. He says, "By the sixties, the chainsaw was how we cut our way through the bush, rather than with axes and hand saws. But it was still extremely hard work. The line cutters were special guys – very tough. At the time, all you had to do to find them was go into the town near your job site and go into a beer parlor. We paid fairly well, so usually within an hour you'd find the people you wanted. Later, I kept a couple of cutters on permanent staff. I want to stress that these were not just labourers; they had to be very skillful, know how to work safely, and also have a bit of sense of adventure."

It is a point of pride for Talson that only once during his career were any of his crewmen seriously injured. "In the late 1960s, I think it was, one of our line cutters sliced himself badly in the leg with a chainsaw. I wasn't in the field at the time, but the guy in charge, the party chief, Rick Sewell, had the presence of mind to bind up the guy's wound to stop the bleeding. Rick drove to the airport in Timmins, hired a helicopter and flew to the site to bring this guy out. Because of his quick thinking, the guy's leg was spared – in fact, he was back at work in a few days."

Talson's retirement party was held in Cochrane in 1999. Invariably modest and good-humoured, he says, "I generally claim that I had to hold the party while I was still signing the cheques, or no one would have come."

Talson's son, Eric, is now a surveyor in Kenora where he runs the firm Rugged Geomatics Inc.

***Talson Rody** is one of Ontario's most experienced and respected surveyors. Following high school in North Bay during the 1950s, he worked on survey crews in various parts of the north, eventually making his way to Cochrane, where the Trans-Canada Pipeline was being built in 1958. He articled to A.M. Mackay in that town during the early 1960s, and took over the business in 1964.*

Above: Muskeg tractors were indispensible to bush travel in the north during the mid-20th century. But often they succumbed to mishaps in the bogs they were designed to cross.

Right: Talson and his beloved Cessna, a companion on numerous surveying adventures in the north.

trum, and undoubtedly from an array of stops between.

Perhaps most importantly, they will, like the best of their clan in any era, do it with their imaginations and ingenuity. Before retiring from their careers, parts of which seem destined to play out in altered climates, at rare latitudes, perhaps even in extended or compressed time, they will see changes not just in the technology and focus of the profession, but in the planet itself – changes that, as surveyors, they will be required to measure and monitor.

The surveyors of tomorrow will see changes that the Erindale grads never dreamt of. Even now, surveyors from the early years at Erindale – men and women familiar with calculus and computer programming and the Lorentz transformations – tell me casually that if they had to go out on a survey site these days they would be profoundly challenged to make use of the high-tech instruments that every young surveyor takes pretty much for granted. In December 2015, a 68-year-old surveyor and engineer told me rather sheepishly that he wasn't sure he was still capable of making a detailed interpretation, perhaps any interpretation, of a contemporary engineering survey. In fairness, it must be said that the standard graphing of a road bed or hydro dam these days is less likely to be a piece of paper bearing a surveyor's signature and a set of ruled lines than to be a screen image, an aggregation of pixels or laser points, in neon colours, comprising a gazillion micro-units of data. At maximum complexity, such an image is likely to be an endlessly modifiable construct, viewable in 360 degrees either from the illusion of great distance or magnified to the point where it strikes the eye in all but molecular detail.

Not long ago, I was reminded by an article in *Scientific American* magazine that such an image will in the near future not even require a screen (perhaps doesn't require one now) but will somehow simply float in the middle of a room, or in mid-air, in 3D, like a hologram, as functional as a topographic map and, at the same time, as elusive as a dream. When I mentioned such images to the Ottawa surveyor James Ferguson, a technological virtuoso renowned for his inovations in GPS and LiDAR (and about whom we will hear more on subsequent pages), he thought for a few seconds, said that he could imagine such a thing, then as if already advancing into some level of sorcerous speculation, said very quietly, "I wonder how they're going to make that happen."

Given the speed at which the future is rocketing toward us, those we are calling "the surveyors of the future" are of course inseparable from the progressive young surveyors of the present. During the past few months, I have met some of the brightest and most enthusiastic of them – talented young men and women

The commissioning of Michael Griffiths (above) in 2002 marked the end of his long journey to become a surveyor in Ontario. Another long journey was completed in February, 1976, when Vladimir Krcmar (opposite page, far right) received his commission from then-Minister of Natural Resources Leo Bernier, and surveyor Al Allman (centre).

whose first language is not necessarily English or French and whose birthplace is not necessarily Canada. Those from, say, Asia and Africa and the Caribbean are, it should be noted, not the first Ontario land surveyors to arrive from afar during the past half century. The current group of "landed" initiates are following in the footsteps of experienced surveyors such as Michael Griffiths, who arrived from Jamaica in 1980, and Tham Shanmugarajah, who arrived from Sri Lanka in 1984; and Gavin Lawrence, who arrived from South Africa in 2003.

Vladimir Krcmar arrived in 1969 from Croatia, struggled through his English-language surveying exams, and in 1980 founded a small company, Krcmar Surveyors, that is, today, one the province's busiest firms, specializing in condo development and run by Vladimir's daughter, Maja, and his two sons, Saša and Tom, all Ontario Land Surveyors.

"When I arrived in Ontario in the early 1980s," Michael Griffiths said in an interview during the winter of 2015, "there was only one African-Canadian surveyor, maybe two, in the entire AOLS; the one I knew was a guy named Omari Mwinyi, from Nigeria – he worked for the City of Toronto for a long time before starting his own firm. When Tham and I joined the Association about fifteen years ago, we made a total of three or four people from racialized minorities. Now, every time I go to the AGM, I see surveyors from all sorts of different cultures; it's great to see the barriers coming down."

Of the twenty surveyors most recently commissioned by the AOLS, a dozen or more brought their talents to Ontario from beyond Canada's borders. I am referring to, among others, Yuriy Bogdanov from Russia, Jayson Ladines from the Philippines, Dacian Popa from Romania, Juan Diego Zapata from Colombia, and Mansour Ghofrani from Iran.

One morning in June 2015, I was summoned from the library at the headquarters of the AOLS, in Scarborough, to be present at Mansour Ghofrani's commissioning. He is a graduate in land survey engineering from Azad University in Tehran, and when I entered the boardroom, where the ceremony was to take place, he was standing at one end of the long table – clean-shaven, impeccably groomed and dressed, his wife smiling beside him, not quite sure what to make of the fuss. He had come an enormous distance to be there, in both the literal and metaphoric sense, and when the executive director, Blain Martin, had delivered a brief and exuberant welcome, Bill Buck, the AOLS registrar, stepped forward and said to this well-travelled young man, "I want you to repeat after me" – and Bill spoke the lines that every new surveyor must repeat after Bill (and try for the next 30 years to live up to): *I Mansour Ghofrani do solemnly swear that I will faithfully discharge the duties of a commissioned member of the Association of Ontario Land Surveyors, according to law, without favour, affection or partiality.*

And with that, Mansour became OLS #1984.

Below: Jansky Lau, OLS #1953, applies that most basic of surveying instruments, the hand shovel, to his endeavours in Toronto in 2016. Opposite page: Surveyor, Andy Shelp, in 2014, at the 50th anniversary of the Ottawa firm, Annis O'Sullivan Vollebekk Ltd., which in 2016 joined forces with Tulloch Geomatics Inc.

"As a profession, we're experiencing a sort of crucial paradox," said the straight-talking Ottawa surveyor, Andy Shelp, in February 2015. "On one hand, we're all about the past, dependent on boundaries laid down 200 years ago, on historical documentation, on procedures that were done or not done, or in some cases should have been done differently. On the other hand, we've come to a point where surveying in Ontario is *entirely* about the future, and will thrive *entirely* on its capability to adapt to change…. More than anything right now, the profession needs the genius and ideas of the immigrant population – let me broaden that; we need imaginative young surveyors, period, no matter what their background."

Of the young surveyors I have met, four crossed my path during a brief period in the winter and spring of 2015 – all of them self-assured, all tech-savvy, all coursing to meet the challenges of their emerging careers. Two of them, Simon Kasprzak and Mel Truchon, were raised in rural or small-town Ontario, while the others, Jansky Lau and Boney Cherian, were raised, respectively, in Hong Kong and in Kuwait. While all four surpassed the brisk educational challenges necessary to becoming Ontario Land Surveyors, Jansky and Boney overcame more specific obstacles in gaining their commissions. A decade ago, for example, Jansky found himself shouldering an increasingly heavy financial burden while enrolled in civil engineering at Ryerson University. During the autumn of 2005, one of his professors, Dr. Mike Chapman, responded to Jansky's financial situation by giving him a job as a researcher in the engineering department at the university.

"During the spring," Jansky said in a recent interview, "Dr. Chapman helped me again by connecting me with the Association of Ontario Land Surveyors, where I worked for a summer in the survey review department."

Jansky's first day at the AOLS has become a fabled memory among his friends at the Association office. When Dave Churchmuch, who at the time conducted the Association's field reviews, handed Jansky a shovel during his first fateful hours on the job and instructed him to do some necessary digging, Jansky was dumbfounded. "Hong Kong is completely urban, no countryside," he told me. "Growing up there, all I knew was concrete, glass and steel. I didn't have a *clue* what a shovel was – had never seen one, couldn't imagine how to use such a thing."

Happily, with a little instruction, Jansky was soon maneuvering the long-handled implement with authority and, in his words, "learned so many other things that summer, and enjoyed the work so much, especially being outdoors, that I decided *this* is what I wanted to do with my life."

Jansky, now a partner at Frontop Surveying Inc., in Markham, acknowledges that he chose geomatics at Ryerson in order to differentiate himself from his engineering classmates, all of whom seemed to be going in the same academic direction. Likewise, Boney Cherian, who attended York University, chose geomatics as a kind of counterpunch to the medical, legal and financial professions that had been pitched to him since childhood as preferred occupations. "Those are the jobs that count in the Kerala area of southern India, where my family lived for generations before moving to Kuwait," he said in February 2015. "My dad is an accountant, my mother a nurse. So they were quite nervous for me when I chose to go into geomatics."

Boney's earliest exposure to surveying, unlike Jansky's, was memorably positive (no shovels). "In our second year at York, we started using a transit and steel tape, doing leveling, transferring elevations, that sort of thing. Then in April of that year, we went up to Algonquin Park

Opposite page: Simon Kasprzak, OLS #1985, canoeing in Algonquin Park shortly before receiving his commission in 2015. Below: Boney Cherian, who surmounted a range of challenges to become OLS #1962, in 2013.

for three days to do a topographic survey. I'd never been to such a place. Coming from Kuwait, I found it exciting just being there in the wilderness, by the water. It was kind of what we always thought Canada should be. When we'd gathered our data, our professor made us draw the contour lines by hand. Back in Toronto they showed us how to create the lines by computer. I just loved it – the whole experience: the park; gathering the data; making the map. I'd always been fascinated by maps – now I knew how to make one! It was as if a little bell sounded, and from that point on I wanted to be a surveyor. When I met [AOLS registrar] Bill Buck for the first time at the geomatics picnic in maybe 2005, he said to me, 'If you want to do it, do it! Become a surveyor! We're behind you.'"

Boney's progress was not always straightforward. "After our third year, everybody wanted to get summer jobs with a certain large firm that hires lots of surveyors. So I applied there, too, but didn't get in." Boney did find work with an alternate firm, which he acknowledges now was not the right place for him. An injury on the job led to a few days off and finally a rebuke from a company official who told him his services were no longer needed. "This person also told me that I'd probably never make it as a surveyor, especially with no family in the profession – that I didn't have it in me."

Boney acknowledges that some surveyors might, at that point, have cut their losses and gone into another line of endeavor. "Instead I got stubborn," he says. "I knew very well that I had it in me. I loved surveying. I had good marks, good guidance. Maybe I should be thanking those people for adding to my determination."

Boney subsequently finished his fourth year, then a fifth in order to get GIS certification. "Just before I graduated I asked David Searles, the Mississauga surveyor, if his company had a job for me, and David said, 'Sure, we do.' And that was that. I articled there, learned on the job – was in the game. I've been a cadastral surveyor with the firm since 2013."

On a fiery-cold evening in January 2015, I found my way to Biagio's Italian restaurant, in the west end of Ottawa, for a gathering of the Eastern Regional Group of Ontario Land Surveyors. That is where I met Simon Kasprzak, who was seated beside his dad, the Renfrew surveyor, Adam Kasprzak, with whom Simon was at the time finishing up his articling. The event that night was an amiable, old-pals' affair at which, for a crucial twenty minutes after dinner, a handful of those in attendance jousted at one another, sometimes quite aggressively, over what they perceived to be the future of the profession. A plainspoken young man, a specialist in Geographic Information Systems at the National Department of Defense, told the 60-odd surveyors in attendance that land surveyors could, if they chose, cling to their traditional interests, shelter within their comfort zone, and steer clear of the latest in data analysis and advanced geomatics. However, they did so at their peril, he said, because if they didn't hitch their wagon to the progressive new world of high technology and complex data evaluation, the data wonks would simply move along without them, leaving them

In her role as Sergeant-at-Arms at the Association's 2013 AGM in Toronto, surveyor Mel Truchon portrayed the movie and video game heroine Lara Croft.

irrelevant to the future. He capped this prickly assessment with the words: "You need us more than we need you."

In an interview later, Simon said, "He made some good points. There are probably quite a number of senior surveyors, well along in their careers, who aren't going to be willing to make the sort of transition that the guy was talking about. Some of them simply don't have enough working years ahead of them."

As for younger surveyors, Simon said, "I haven't heard about a single one, anywhere, who's reluctant to embrace the future – or get involved with the more conceptual aspects of the profession."

He added that, beyond these complexities, there is also a range of "more basic challenges" facing young land surveyors as they move forward. As an example, he noted the increasing availability of survey-grade GPS, at ever-lower costs, and of the extensive cadastral data available on the internet. "It won't be long before a guy with a bit of knowledge and a decent GPS unit can survey his own property."

Simon pointed out, too, that some 60 percent of the active surveyors in the province are in a demographic that is set to retire within half a dozen years. "When they do," he said, "it'll all be up to the young folks – not just to operate the firms but to run and regulate the profession; it's a lot of responsibility."

On the bright side, as Simon sees it, the smaller number of surveyors is going to "throw the profession wide open" – to make it possible for a newcomer to "take pretty much any route or role he or she wants. Which for me is a very exciting prospect. Certainly, I wouldn't want to be doing anything else right now."

Meanwhile, it should be noted that at the time of the interview, Simon (and just about everybody else in the profession) was unaware of a striking development that came to light in early 2016. Compared to the 30-odd surveying students articling in the province five years ago, there is today, according to AOLS Deputy Registrar, Maureen Mountjoy, a cohort of more than 80 students articling, the highest number in decades (this thanks largely to the concerted efforts of the Association to promote surveying as a profession). So the drop-off in numbers is unlikely to be as pronounced as might have been anticipated.

As Simon, Jansky, and Boney have worked their way through the early thickets of their careers, Mel Truchon has had nearly nine years of favourably condensed experience at Tulloch Geomatics, one of Ontario's most progressive survey and engineering companies. Raised on a Peterborough-area farm and commissioned in 2011, Mel has, despite her youth (perhaps equally because of it), been put in charge of engineering surveys at the ever-evolving firm.

With 275 employees in ten offices across the province, Tullochs is headquartered in a striking new building in Huntsville, Ontario, where I fetched up one morning in May 2015 to interview the company's president, Mark Tulloch. A few months earlier, a senior surveyor from Ottawa had told me that Mark is one of a handful of identifiable visionaries in contemporary surveying in Ontario. It was on a walkabout after our interview that Mark introduced me to Mel, who at that moment was ensconced in a kind of boardroom, listening intently to a representative of a high-tech instrument firm pitching her on a new FARO terrestrial LiDAR scanner. (As contemporary surveyors are aware, LiDAR, whose name is an acronym for Light Detection and Ranging, functions by sending out hundreds of thousands of laser beams a second, collecting them again as they rebound off their target, and arranging them, via computer, into a 3D image of whatever is being scanned, from highways to mine tailings to the interior of buildings.)

Mel had in recent months orchestrated an array of mobile LiDAR surveys, assessing among other things freeways and bridges, as well as one of the world's longest and largest structural culverts – a conduit that channels water from the Niagara River, above Niagara Falls, to the Sir Adam Beck hydro facility downstream. "We were basically checking for deformation," she said, adding that conditions were "a little different" down there – "actually pretty exciting; it was quite an adventure. You work in a fall-arrest harness, on a narrow catwalk, more or less in the dark, above millions of tons of water just roaring along through this cavern below you."

Given the epic dimensions of such work, engineering surveys are, at the same time, things of extraordinary intricacy. Three years ago, Tullochs did an engineering survey of the runways at Ottawa Airport, where, as Mark noted, "We were looking for the tiniest imperfections in the surface, which under certain conditions can create havoc for incoming jets. The LiDAR unit can pick up even the most minute cracks and pock marks – imperfections very difficult to see with the human eye."

Mark explained that a similar sense of micro-investigation is central to the firm's ongoing surveys of the massive steel structures that support overhead signs on multi-lane highways. "In the past," he said, "we'd have had to do those surveys by lurching along from one sign to the next in our trucks, getting up onto these structures with cameras, having our people working in vulnerable, even dangerous, situations, as the traffic whizzed by." Today the company does them from a vehicle moving at highway speed, using remote photogrammetric methods, capturing visual data that the company's engineers assess on computer screens in the safety of the office. The captured images are so detailed that even a smear of corrosion, hairline fracture or unstable weld is exposed.

Meanwhile, with her ever-expanding resumé, Mel has become something of an avatar for what is possible for a young surveyor in the right place at the right time, in the right firm, in 21st-century Ontario.

Mark said, "We more or less told Mel, 'If you want to lead our engineering survey group, do it. We'll support you.' It's the sort of thinking that bites you at times, because everyone has to learn, and that can mean mistakes. But too often I've seen the older generation cling to control, unwilling to pass the torch to bright and energetic young surveyors. So the young people just stagnate – or quietly drift away. If you give them the right amount of responsibility – basically, as much as they feel they can handle – they thrive, and so do we. And so does the profession as a whole."

During the months that followed, I met other young surveyors, including Sophie-Rose Côté, from the Fredericton area – newly commissioned as an OLS but already with a continental resume that includes work on the Trans-Labrador Highway; on Highway #11-17 north of Lake Superior; on topographic surveys in the remote Peace River region of British Columbia; and in preparatory surveys for the installation of wind turbines on the eastern shores of Lake Huron.

And I met Vaitheki "Vicky" Kumar, who arrived in Canada as a child from

Family Matters
(and nowhere more so than in Ontario surveying)

During the 125-year history of the Association of Ontario Land Surveyors, just 2,000 men and women have earned the right to place the letters OLS after their names. In this relatively compact group, more than sixty families have produced two or more surveyors, and in some cases, three, four or five surveyors. They are, among others, the van Nostrands, the Bairds, the Keatleys... the Aylsworths and Stidwills and Monteiths ... the Hileys and Bunkers and Simpsons... the Rodys, Meldrums and Purcells. They are fathers and sons; fathers and daughters; brothers and sisters; husbands and wives. In some cases surveyors from two generations of a family have each served as AOLS president – the Andersons and Kirkups and Brubachers among them. Herewith, a tribute in photos to a number of the surveying families of Ontario.

1

Peter Gibson (commissioned in 1858) and his wife, Eliza, seen here in 1901 surrounded by two generations of their offspring including sons Harold, Wilbert and Morton (all surveyors), grandsons Ansel and Silas (who would become surveyors), and four daughters.

1. *John Hiley, Sr., and John Hiley, Jr., canoe-racing; circa 1980.* **2.** *Vladimir Krcmar with daughter, Maja, and sons, Tom and Saša.*

Opposite page: **3.** *Norm Sutherland with sons (left to right) John (a Michigan surveyor), Bloss (OLS) and Greg.*
4. *Dave and Wayne Brubacher.*
5. *Bob and Maureen Mountjoy.*
6. *Julia and Hazen Meldrum.*

2

3

4

5

7. Mike Tulloch with sons Dave (left) and Mark. 8. Tham Shanmugarajah with wife, Kumu, and sons, Hari (right) and Rocky who is currently articled to his dad. 9. Jim Purcell (left) and son, Murray. 10. Vaitheki "Vicky" Kumar at her commissioning (2015), with surveying dad, Chandrasekarem, and family.

10

6

7

9

8

Sophie-Rose Côté, OLS #1987, is seen here before expectant fans, participating in the two-person crosscut event during her years as a member of the University of New Brunswick's Woodsman Team.

Sri Lanka, via Dubai, in 1995. She got her first exposure to surveying as a teenager by working summers for her dad, Chandrasekarem Kumar, who had worked as a surveyor in Sri Lanka and the Middle East and had found work in Canada as an affiliate of P. Salna Ltd., a surveying firm in Richmond Hill. Vicky attended Sir Wilfred Laurier University in Waterloo, hoping to become a geography and math teacher. But given the poor prospects for teaching jobs when she graduated, she began working full-time with her father. "To be honest, I hated it," she said. "The hard hat, the safety boots, the muddy suburbs, the cold. Plus, the only thing we were doing was residential work, subdivisions. But gradually I got the hang of it, developed a bit of a knack, and was able to provide some real help to my dad. I'd have everything prepared when we got to a site; he told me he'd never had anyone do this for him. Then one day he said, 'Why don't you get your credentials and become a surveyor?'"

Which is exactly what Vicky did – brushed up on her academic qualifications at York University and, after some soul-searching, cut working ties (amicably) with her dad, in order to seek a broader experience of surveying than she could get laying out residential lots. Her next ambitious step was through the wide glass doors of one of the province's biggest and most diverse firms, J.D. Barnes Limited, where she took up articling in 2013, and where she is now employed as a fully fledged surveyor. "It's funny," Vicky said in March 2016, "You begin your academic courses to become a surveyor, and you think, okay, my education in this field has begun at last. Then you start articling and you think, okay, my education in this field has begun at last again. Then you join the real world of surveying, and you think, okay, my education in this field has begun at last *again*… And with every new project, again."

If our focus so far has been change, it might equally have been what has *not* changed in the world of surveying in this province during the years since the lions of the past – Willis Chipman, Alexander Niven, Thomas Speight, *et al* – put their heads and ambitions together and founded the AOLS in 1892. There is a temptation while attempting to put words to the enduring character of the profession to trundle out the usual suspects – abstract nouns such as "honour" and "integrity" and "commitment." And indeed such qualities are the lubricants on which surveying or any other profes-

Julia Meldrum Smith, of Cornwall, followed her dad, Hazen Meldrum, into surveying during the mid-1990s. Hazen's equipment and artifacts, which Julia displays in her office, are a clear and nostalgic testimony to the changes that have taken place in surveying during the past half century.

sion is likely to thrive.

However, where a good many surveyors are concerned we are talking equally about more earthly graces: about forthrightness and tenacity and a sense of survival and adventure. We are also talking about a healthy variety of vulnerabilities – exposures of the sort that tend to humanize a profession and, in a good many of the surveyors I have met, to show up as variations of modesty and an absence of pretension. The first surveyor I ever knew was Hazen Meldrum, OLS #959, an unassuming Maritimer, who practiced in Cornwall, Ontario, where I spent seven years as an adolescent and teenager. When I interviewed his daughter, another surveyor, Julia Meldrum Smith, in May 2015, in Cornwall, she said that when she took over her dad's practice as he eased into retirement, folks from the area, some of whom hadn't seen Hazen in years, would sometimes say to her, "If you're anything like your dad, we're going to get along fine." In other words, people still identified Hazen primarily by his modesty, plain-spokenness and sincerity – and presumed that his daughter would possess those same qualities (partly, of course, because she is his daughter, but partly, too, in this world of slanted logic, because she is his daughter *and* a surveyor – *isn't that what surveyors are like?*).

In March 2015, at a social function, a senior Ontario surveyor whom I have heard described both as "acerbic" and "esteemed" stepped away from a group of surveyors with whom he was conversing and said to me in a stage whisper, "Don't ever quote me on this, but one of the things I've observed about our kind over the years is that we're a bit like black bears – we don't always expect to get our own way. It keeps us from getting uppity."

As he rejoined his group, he said over his shoulder, "I've met folks in some professions who remind me more of grizzly bears." This brought a cartoon growl from another surveyor and a suggestion that a few grizzlies may have jumped the fence recently and were cross-fertilizing certain black bears.

Surveyors are unquestionably feeling some very real vulnerabilities these days. Beyond the challenges noted by Simon Kasprzak, there is a pocket of skeptics who think the profession may someday be swallowed up in rarified methodology or absorbed into the proliferating advance of engineering; or will get so compact that it will lose its critical mass and thereby its treasured status as a self-governing profession, its right and obligation to regulate and discipline itself.

While indeed small in size, surveying as a profession has, as they say in boxing, always fought above its weight. The articling student James Dorland told me recently how surprised he had been, on visiting a restored pioneer village, to find that "the surveyor's house was bigger than the doctor's or anybody else's."

"We're durable; we're cohesive," says Blain Martin, part of whose job as executive director of the AOLS is simultane-

During a 40-year career, Izaak de Rijcke, OLS #1458, has been a surveyor, a lawyer, a university instructor, a watercolour painter and a writer... and has focused all of those activities on his interest in boundaries and the land.

ously to limit and expand the profession's relentless inquest into its own health, stature and image.

At the moment, there are roughly 500 surveyors practicing in Ontario, compared to some 30,000 medical doctors, more than 40,000 lawyers, and 115,000 elementary and high school teachers.

In the view of the former executive director, Lorraine Setterington, the Association has been successful not in spite of being so small but rather because of it. She says, "I went to Copper Cliff high school; we had two hundred and fifty kids. We won all-Ontario hockey twice while I was there. We had spirit. It's the same at the Association."

Martin says, "It's a pretty good indication of our *esprit de corps* that, while most professions get less than a twenty percent turnout at their annual meetings and conventions, we seldom get less than seventy-five percent."

It has been suggested that the sense of occupational vulnerability is perhaps also a function of a more generalized concern, common to many professions, about the uncertainties of the future itself. York University's Dr. Matthew Johnson, who refers to himself as "an early universe cosmologist," might well have been talking about surveyors when he said that processes and technologies that were unknown six months ago "appear in our midst, enter our work lives, become functional, become routine – are sometimes even transformative – and are just as quickly obsolete, displaced by potentialities that are themselves expanding constantly into higher, wider orbits."

The Guelph surveyor and lawyer, Izaak de Rijcke, an inspired fount of perspectives on where surveying has been and where it might be going, observed

recently that at the time of the formation of the AOLS, 125 years ago, surveyors "were a significant, even vital, resource" in their communities – knowledgeable men who laid down boundaries and knew where others had laid them down, and were not about to be second-guessed on that knowledge.

"They were the experts," Izaak said. "They had well-deserved status in the society of the day, and they embraced that status – built on it."

The problem today, as Izaak sees it, is that surveyors tend to make the same assumption about their stature in contemporary society. "But it doesn't work anymore," he shrugged. "Society has changed. Institutions have changed. In their rendering of boundary decisions, the courts have demonstrated that boundaries no longer need to be laid out by a surveyor."

Increasingly, according to Izaak, information on boundaries comes from lay people, men and women who know more about community land use – access lanes, drainage mechanisms, shorelines – than any outsider with a GPS unit or town plan. "As surveyors, we should be listening to those people, asking them what they can tell us, and in turn asking what we can offer them. It is just so important right now to be looking outside of our own circle – that's where the discoveries are going to be made."

Surveyors, concludes Izaak, "are still experts in *interpreting*, in managing information of the sort that, historically, came to us from geologists, foresters, botanists, all manner of geographers. It was that multidimensional aspect of a surveyor's training and knowledge that kept him relevant. That's where we want to be again."

This high-quality brass transit was manufactured by Charles Potter in Toronto during the mid-late 1800s, when most surveying equipment was coming into Canada from established manufacturers in the U.S., England and Germany. Potter also manufactured telescopes, opera glasses and spectacles.

On a snowy day last March, I had the privilege and pleasure of sitting in the office of Ontario's Surveyor General, Susan MacGregor, in Peterborough, chatting with her about the state of the profession's past, present and prospects. For now, suffice it to say that one of her concerns that day was what she called "the public perception of surveyors," or more precisely the absence of a public perception – "Hardly anybody knows who we are!" she lamented at one point. "People know *so little* about all the vital things we do!"

My own understanding of those "vital things" runs like a nerve between two transformative moments, the first of which (inspired by my earliest inquiries into surveying) revealed to me the vast and lasting contribution that surveyors have made to the history and evolution of Ontario. There is no limit to the significance of establishing boundaries in a culture whose foundational ideal was and remains the right to own or control land, to mortgage it, develop it, farm it, build on it, play on it, pray on it, dig in it – to make a living or, in some cases, a fortune from its bounty. As we are daily finding out, there is also, sadly, no limit to the significance of scalping it, gutting it, poisoning it – in the worst cases reducing once-rich soil to dust, and once-clean waters to sewage or worse... all of which must also be surveyed.

Even today, the spidery old boundaries laid down by the surveyors of history knit up the planet as surely as a fascination with the planet knits up the surveyors. During a 63-day walk from my home in Thunder Bay to New York City, in 2001, I got an inescapable look at the profusion of Ontario's boundaries – and an equally inescapable look at everything that *cannot* be divided or bound: the intricate continuity of forests, mountains, beaches, water, air – the endless interlacing of habitats and species ranges; sunlight and soil; streams making lakes making clouds... making rain and maple sap and moose blood. It has occurred to me several times during the writing of this book that unless one recognizes and appreciates the intrinsic oneness of what has been divided (as I believe surveyors do), it is impossible to understand with any depth the *fragmentation* of that oneness into districts, townships, counties, municipalities, police precincts, political ridings, conservation zones, Native land, Crown land, parkland, private land, everything down to adopt-a-highway and snowplow zones: the world as a jurisdictional game board, all stops and boundaries more obsessively championed than is the indivisible world of nature – or of the "common ground" that was present-day Ontario prior to European arrival some 400 years ago.

> ***Unless one recognizes and appreciates the intrinsic oneness of what has been divided (as I believe surveyors do), it is impossible to understand with any depth the fragmentation of that oneness into districts, townships, municipalities ...***

Our historic surveyors and planners were anything but infallible in seeing the relation of the natural and the divided world. Norman Wilson, who was president of the AOLS in 1943, deplored, for example, how Toronto's 19th-century lot lines dictated the street system, which he pointed out ran "diagonal to the topography," resulting in "enormous additional costs in city development." He loathed, too, that, out of limited vision, early surveyors had laid out the province's 100-acre farms in such a way that some included 90 or more arable acres, while others included just

50 or 60. More radical thinkers such as the great American land ethicist, Aldo Leopold, believed that the culture's entire reduction of "land" to "real estate," via circumscription and ownership, was a debasement of the integrity of the planet. He believed further that governments were the least responsible land owners, because of their ever-shifting values and corporate cronyism, and that it was private citizens, those with a sense of protection of their home soil, or with conservationist ideals, who were the least likely to destroy planet Earth.

To EXPLAIN the second of my transformative moments, I must first note that while writing these chapters I frequently worked late into the night, alone, at the AOLS offices on McNicoll Avenue in northeast Toronto. Those offices exist in a presentable but unremarkable office park, next to a vast and unremarkable parking lot, in a remarkably unremarkable suburb sometimes described jokingly as Scarberia. There, at maybe 3 a.m., on a snowy night in February 2015, perhaps six weeks after I had begun work on the book, I got up from my desk in the office's second-floor library, sauntered down the hall to the much larger boardroom, and laid out a map on the giant table that pretty much fills the room.

Twenty minutes later, on a break, as I stared out the window through a scrim of falling snow, it dawned on me that everything of substance within view had begun with a surveyor: the street, the sidewalks, the curbs, the street lighting, the parking lot, the other buildings in the office park, our own building, the pipes delivering gas to all these buildings, the lots on which the buildings sat, the landscaping, the houses in the distance, the apartment complexes, the massive hydro towers and lines that cut the sky into triangles and scimitars; the power dams and reactors and gas-fueled generators that somewhere out yonder were feeding their endless supply of electrical power into the nearby wires. I cannot recall whether, as I stood at the window, I saw any of the hundreds of jetliners that fly over the Association office on their way into or out of Pearson in the course of a day; but it is worth noting that even they depend on surveyors – not just for the making and maintenance of airport runways but for every detail of a jet's flight plan, as described to me one day by the Midland surveyor Dale Eplett, who has worked on such plans. The sky itself, with its stars and planets, was for centuries one of the surveyor's closest working companions.

The gist of that memorable moment was my recognition that surveyors, the amiable and intelligent tribe with whom I had been at work for the past six weeks, had been present not just at the shaping of Ontario but are *still* fundamental to almost every residential, industrial or infrastructure development that occurs within the province's borders. Walk up Yonge Street or along a trail in Algonquin Park; drive the north shore of Lake Superior; climb the CN Tower; sit in the Rogers Centre; board the train in Cochrane; run a bath in the Chateau Laurier; take the escalator at the Thunder Bay Airport; emerge from an elevator into your 30th-floor condo; shoot a puck in the Meridian Centre in St.

The Toronto firm of Marshall Macklin Monaghan played a significant role in the surveying and engineering of the CN Tower, completed in 1976. Numerous land surveyors participated in the tower's construction.

In the summer of 2005, the Espanola surveyor Bob Halliday travelled to Nunavut to place permanent markers on mineral claims that had been staked along the Arctic Circle. The photo below, entitled "Ice Wall," shows Bob's assistant, Kyle Ollerhead, at work 100 kilometres east of Repulse Bay, along the Northwest Passage.

Catharines. Do any of these things, and know that they are possible because at some point in the near or distant past a surveyor has made a precise assessment of the surroundings and has determined, say, that *this* is where the highway might best run in order to provide the safest driving or the solidest or straightest road-bed; or that this is where the power dam or hydro towers should be constructed; or where the piers must be located to properly support the bridge. Or that this, Madam Developer, is how to wring two hundred and eighty building lots and a skein of streets and sewers out of what used to be a couple of rural farms – and before that was a stand of old-growth pines; and before that a hangout for wooly mammoths; and before that a sheet of ice some 400 metres thick.

Blain Martin told me on the day we met, in late 2014, that he had once worked for NAVTEQ, a company whose efforts in digital mapping were a precursor to online maps and GPS, which, understandably, were developed around the work of many surveyors. He further explained that MRI scans (in effect a kind of 3D mapping of the interior of the body) are also tied to surveying. As are all manner of terrain- and space-plotting systems, and military defence plans, and geophysical assessments; and the monitoring of oil spills or the tracing of industrial toxins through river systems or deep into the ground. A while back, on Lake Erie, surveyors worked offshore to determine the precise location of a downed airplane beneath shifting plates of ice. Surveyors are in fact early on the scene at the site of every plane crash in Canada, attempting to determine, among other things, flight vectors and the spread and pattern of debris ("It's one of those jobs that surveyors never really want to do," said the Ottawa surveyor, Dave Woodland, in 2015). Meanwhile, in Ottawa, a hospital employed surveyors to map the spread of germs and disease through its wards. In London, anthropologists teamed with surveyors to create a surface map of the human body, indicating the places where people were most tolerant or intolerant of being touched.

The survey technologist, Pat Hills, a Toronto segment manager for the Canadian survey equipment firm, Cansel, said in October 2015, that he is regularly consulted by surveyors and engineers on issues as diverse as, say, the stability of the towers on which contemporary roller-coasters, such as the Leviathan and Behemoth at Canada's Wonderland, rise and plummet; or the method and equipment that might best be used to measure the sway of an office tower or residential highrise; or whether or not an overpass on, say, the 407 is shifting (if not, why did that chunk of concrete fall from it?); or the degree of glacial rebound in Ontario, where the rock mantle beneath

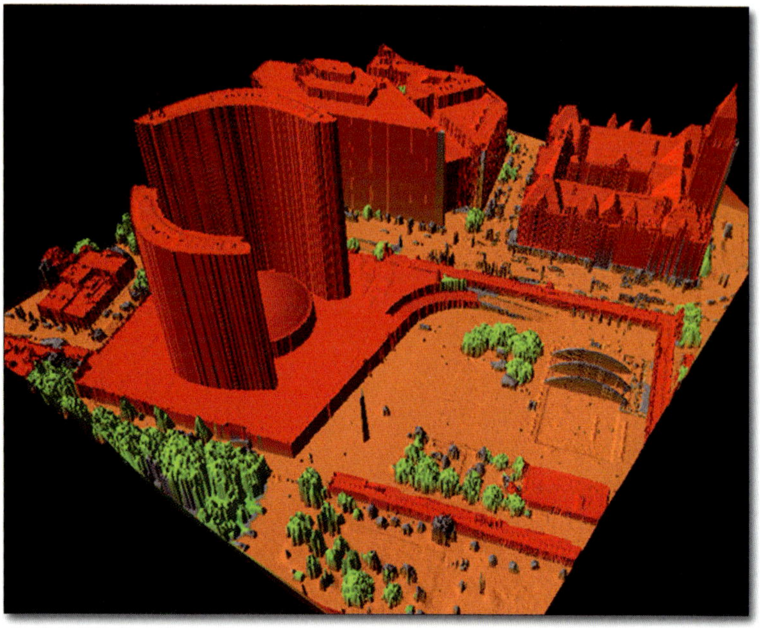

The urban and the rural. At left, a LiDAR scan of Toronto's new and old city halls. Opposite: The human shaping of nature, as seen in the foreground orchard, pitched against the broader order of the spruce forest, outside Thunder Bay.

and around the Great Lakes is still rising – bouncing back – some 12,000 years after being freed of the weight of the Wisconsin Glacier.

On a day last winter, the surveyor Michael Griffiths explained to me that he and his one-time university lecturer, Ron Stewart, were strategically involved in calculating how much land was lost by several First Nations reserves on Rainy Lake and Lake of the Woods, when, decades ago, the provincial government built dams to raise water levels in order to facilitate navigation and the floating of timber to the mills. Such calculations would determine how much compensation might be due the local bands these many years later.

Maybe a week after I chatted with Susan MacGregor about the relative anonymity of the surveyor, another surveyor said to me on the same theme: "For all we do, when do you *ever* hear a news story, or *any* media story, about a surveyor?" He pointed out that there have been novels, TV shows, movies, even comic books, featuring every profession and occupation known to humanity. He rattled off a litany of them, beginning with "the oldest profession" and carrying on through teachers, farmers, journalists, ad-men, oil men, ranchers, doctors, pilots, stewardesses, lawyers, sea captains, architects, policemen, judges, taxi drivers, veterinarians, actors, pawnbrokers, restaurateurs, writers, dancers, fishermen, junk liquidators, athletes, mathematicians, stock brokers, Mafiosi, funeral directors... His chagrin over the (arguably merciful) exclusion of surveyors from pop culture reached its apogee in the observation that there is "now even a reality show called 'The Librarians!'" He was certain, meanwhile, that there would "*never* be such a series about surveyors" – that surveyors were destined to remain shadow operatives, as mysterious as Garbo, in the grand pattern of pop and media culture.

We agreed that a historic documentary on famous surveyors might fly – featuring the lives of, say, David Thompson, George Washington, Abraham Lincoln, Thomas Jefferson, Sir Sandford Fleming, or the writer and philosopher Henry David Thoreau – all of whom, as the joke goes, knew the length of a 66-foot chain.

Meanwhile, in the shadow of the surveyors' anonymity, Susan MacGregor is determined that in 2017, the 125th year of the existence of the AOLS, she and the

surveyors of Ontario will, in her words, "take the basket off our light – let the world know who we are."

If there is a grace note in all of this, it is that the surveyors' implied antidote both to their cultural anonymity and to uncertainties about the future, is the simplest, most eloquent and self-sustaining of humanity's survival mechanisms: they are well grounded.

During the past year I have discovered (and rediscovered) that the characteristic that most links surveyors to one another and to their profession – that most links them to themselves – is a passion for, and curiosity about, the earth. They are people of *place* – not just of water, soil, riverbanks, forests, and the like, but of cities, skyscrapers, highways, bridges, tunnels... of farm lanes and fjords. Those with a less literal affinity for shoreline and rock – let us say a sublimated affinity – tend to attach themselves to the *mathematics* of place: the curvature of the earth, the shifting poles, the azimuth of the star... to geographic information on the very wherefore and why of the physical world.

Almost every surveyor I met during a year of interviews and conversations told me he or she had gravitated to the profession out of some level of fascination with the physical world. If they did not tell me directly, it was evident in their stories and recollections – those of Bill Webster, John Goltz, Jim Hill, Tim Hartley, Andrea Tieman, Bob McDermott, Anne Cole, Helmut Grander, Tom Bunker, Charlie Fairhall, Talson Rody, Simon Kasprzak, Jack Young, Julia Meldrum Smith.

On an afternoon in December 2015, Blain Martin came into the AOLS archive where I was working and said, "One morning when I was a kid on a survey crew in Orillia, I was walking out into the woods just after dawn, with my pack on my back, my axe over my shoulder – heading out to cut line, which is not everybody's idea of a good time. I came to a place where the sunlight was coming through the trees, and I stopped for a moment, and it hit me: the beauty of it, my elation in being there, the prospect of the day ahead, and finally the realization that *this* made me happy, that this is what I wanted to do with my life."

Anne Cole tells of graduating from high school and heading off into the wilderness to an all-but-mythic hideaway called Camp Manitou, where her family had for years been active participants,

In 2016, Bob Clipsham, seen here with his wife Pamela, enjoyed his 46th and final year in surveying. Below: The family's long-time connection to Muskoka and to the land, the Uneeda Rest Hotel, has been in the family since the late 1800s.

near Whitefish Falls, on the north channel of Lake Huron. There, she and a couple of friends spent the winter of 1975-76 in a wood-heated cabin, figuring out what they wanted to do with their lives. Anne says, "I had the academic standing to choose law or engineering but I wanted something more adventurous – wanted to be outside. I'm a paddler, an outdoors person." She did research into surveying which, as Anne puts it, "turned out to be the very thing I was looking for."

Bob Clipsham's first comment to me when we chatted in February 2015 was that he didn't have much to say. And at first he did not – seemingly by choice. But as we warmed to a discussion of our shared connections to Muskoka, which for both of us is an ancestral home, the story emerged of his deep inner commitment to old roads, old rock, old waters and woods, old buildings, including the wooden resort hotel, the Uneeda Rest, built by his grandparents in 1898 on farmland purchased at Kilworthy, on Sparrow Lake, when the family blacksmithing business went into decline. It is a dwelling that the Clipshams and their extended family still occupy in the warm weather. "I go there to hide out from it all," Bob said. "To get back to what I was born to and still love."

Bob also owns the 100-year-old blacksmith shop that his family ran nearby. He possesses blacksmithing skills that he acquired as an adolescent, and maintains an interest both in the gardens that his ancestors planted to survive and in the boats his father and grandfather built on Sparrow Lake. He and his wife Pamela live on the street where Bob's forebears settled in Churchville upon arriving from England in 1847.

Regarding the many young surveyors he has instructed or guided through articling over the years, Bob says, "The ones who adapted most readily to the profession were invariably the ones who were connected in some way to the land. Maybe it was as simple as their family had taken them camping when they were young. Or they'd been raised on farms. They knew what a rush it was after a rainfall when the runoff would increase, and the stream would swell, and the water would get into the ground and filtrate. When they got to school and out into the field, they quickly understood processes such as urban runoff and the hydrological cycle. Plus, they could manage themselves in the bush, could make seasoned observations."

Likewise, Maureen Mountjoy described to me one day, in rhapsodic detail, the simple summer cottage where she and her sister and parents spent summers near Dorset during the 1960s –

William Webster • Ink to Paper to the End of Time

"**Once when I was a kid, in Ottawa,** I saw surveyors measuring the distance to a communications tower by taking the angle to the top of it and combining that with the known height of the thing. As I watched these guys I felt a kind of stirring -- sensed a refreshing power in what they were able to do, and so simply.

"Later, I worked summers as a surveyor. In winter I'd cut classes at the Carleton University and head out with a crew at minus-30 degrees. I loved the science, the math, the law. But there was nothing I loved more than being outside.... A friend once said to me: Paper never refuses ink. When you're in the office you can pour ink to paper, punch down data, to the end of time – in other words, address the technicalities of surveying, the mechanics and numbers.

"Life in the field is what *produces* that data, but it also transcends it, gives you the *poetry* of surveying: I'm talking soil, trees, rivers, stars – maybe a hawk or an owl, or if you're lucky some interesting people. And of course air – fresh air. Even in my late sixties, I try to get out as often as I can – I still love a good crisp day in January or February, especially if there's a problem to solve. That's my specialty – solving legal boundary survey problems. I'll stand at the site, flare my nostrils, take it all in. Then, based on my gut feeling for what needs doing, I'll do it – I'll come up with a solution."

William Webster, OLS #1319, was commissioned as a surveyor in 1972. He currently works with W.J. Johnston Surveying Ltd. in Winchester.

Gord Keatley • Pay to the Order of …

During the early 1970s, Lou Emon, a surveyor on Manitoulin Island, got so busy that he talked William Keatley into leaving his surveying practice in Sudbury and opening an office in Little Current on Manitoulin's north side; they would share the island's work. The arrangement held into the early 1980s when Lou closed his office and, with his long-time assistant, Olric Hembruff, went to work for Bill Keatley.

"One of the distinctions of Lou's operation," said Gord Keatley, Bill's son, in February, 2015, "was that he always worked with this enormous wooly dog, named Ace. You'd see Lou coming down the road in his big old Pontiac LeMans [which Gord eventually drove in a demolition derby], and invariably Ace would be hanging out the passenger-side window. In the field they'd tie orange tape to Ace's tail and collar, so that he wouldn't get mistaken for a bear and get shot. Anyway, long after Lou retired in 1991, I was digging through some files, and I came across this old cheque written in the early 1980s by Lou Emon. I read it – then I re-read it. The thing was made out to Ace the dog – it was for ten dollars and a bag of biscuits.

"I took it to show Dad, who was long retired, and he started laughing. What had happened was Lou and Olric were doing a job down at Providence Bay, on the island; they had to take a measurement that crossed the Mindemoya River, which is about a hundred feet wide. This was in the spring, so there was high run-off, very cold water. The only way the guys had to measure a distance back then was with a 200-foot steel tape. They thought for a few minutes about how to calculate this distance, arrived at a solution, and Olric got in the car and drove way upriver to the bridge and back along the far shore. When he appeared opposite Lou, by the river's edge, Lou tied the end of the steel tape to Ace's collar, and Olric started yelling for Ace from the other side. With no hesitation whatsoever, Ace jumped into the river and swam straight across with the tape.

"The cheque I unearthed many years later was one they'd written over a beer, at the end of the work day, to honour Ace's commitment to the art of surveying. Their joke at the time was that between Lou and Olric they had eighty years of experience on one crew… and another five if you included Ace."

Gord Keatley, OLS #1883, operates Keatley Surveying Ltd., the only survey firm on the world's largest fresh-water island, Manitoulin.

The Mindemoya River, Manitoulin Island.

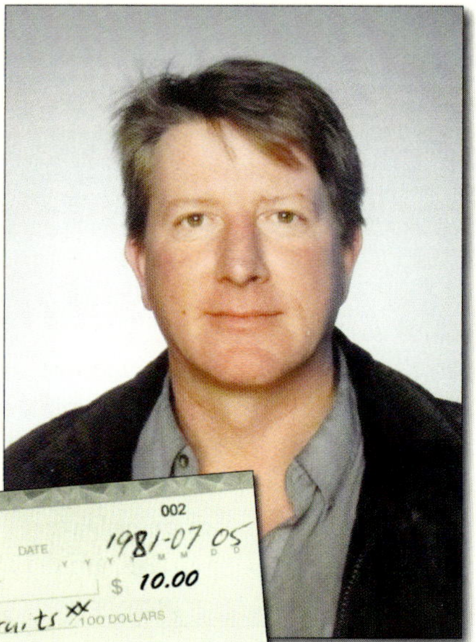

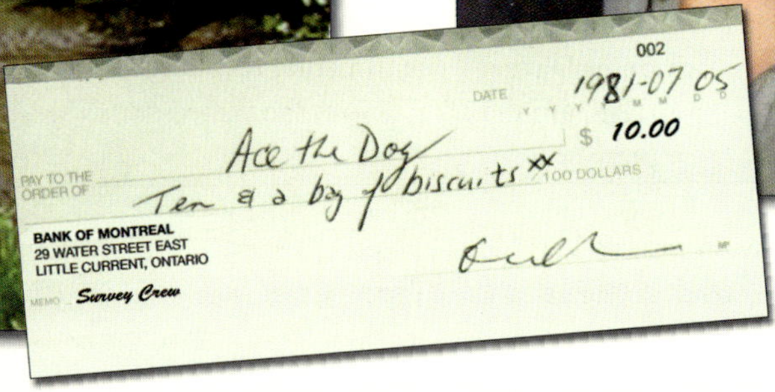

The renowned naturalist, surveyor and professor, David Lambden, during his years at Erindale.

doing construction, carrying water, cutting wood, lighting fires, swimming in the crystalline lake. In subsequent years, she ran the survey camp at Gull Lake in the Minden area. "Deep down, I'm sure it was those early experiences in the woods, by the lake, the whole outdoor dream, that ultimately made surveying attractive to me."

When Ron Stewart paid tribute to the venerable Erindale professor, David Lambden, on the occasion of Lambden's Professional Recognition Award in 2006, he recalled in his testimonial a day on which he and his former professor had gone out together to do a survey in a swampy bush somewhere northwest of Toronto. "I have pictures of David going through swamp water up to his chest, and appearing to enjoy it," Ron said to a crowd of nearly 500 surveyors and their spouses. "When we were walking through the dry part of the bush, he identified all the flora by their Latin names. I had no idea of course whether he was telling me the truth. At one point, he said, 'Listen to that bird – that's a rose-breasted grosbeak.' I said to myself, 'Yeah yeah, sure' – at which point Dave announced, 'And there he is!' He pointed to the bird, and it was in fact a rose-breasted grosbeak."

In accepting the award, Lambden called surveying one of "the professions of the land" and recalled his earliest surveying gig, for Ducks Unlimited, in 1945: "I got the job because of my passion for wildlife; it's been with me all my life. And, yes, I do know most of the Latin names for the fauna and the plants as well."

Bob Gunn, another celebrated professor at Erindale, now in his late eighties, lives with his wife Marie on the 12th floor of a deluxe high-rise in Mississauga, where I visited him in May 2015. The apartment's east-facing windows look down directly onto the wooded Erindale campus where Bob taught a good many of the survey program's 450 graduates and

where Bob still walks regularly. "I love those woods," he told me with a wistfulness that belied his broader sense of unsentimental recollection.

From the same windows, we could see a vast expanse of Lake Ontario. He remarked on an oceangoing vessel leaving Toronto harbour, and we began talking about the St. Lawrence Seaway, in which we both have a stake – I because I grew up in Cornwall, "the Seaway City," Bob because, years ago, he developed a way of gauging the movement of the huge Moses-Saunders Power Dam. The two-million-ton concrete fixture, at Cornwall, is central to the seaway's existence because, quite apart from the dam's capacity to produce hydroelectricity, it raises the water-level upstream to navigable depth. "People think big dams are stationary," Bob said. "But they're not; they move – not much mind you. The cold does it. Then they move back. It all has to be measured." As he spoke, a hawk, with a head the size of a baseball and what must have been a four-foot wingspan, glided out of the sunlight toward the 12th-storey window, veering off at the last second, giving us a thrill and pasting the moment indelibly onto the interior of my skull. "You can't beat the planet for surprises!" Bob enthused. When I was ready to go, he did me the honour of presenting me with a surveying problem, a modest geodetic excursion into the magnificence of the scene

Surveyor and Erindale professor, Bob Gunn, is seen here on Old Crow Mountain, setting up control for aerial photo mapping of the Old Crow River Valley, in the Yukon, in 1977. Lonely Mountain is barely visible in the haze across the valley, while the Porcupine River flows below. Bob, who had been transported to the mountaintop by helicopter, took the photo himself, using the timer on his camera. A Tellurometer model MRA3 microwave distance-measuring device sits to the right of the tripod.

before us. I will not go into detail, except to say that thirty kilometres away we could see the CN Tower and the cluster of skyscrapers that give shape to downtown Toronto. Bob said, "Because of the curvature of the earth, the tower slants slightly away from us – in other words, the top of the tower, the very centre of the top, is further away from us than is the centre point of the bottom. The question is, what do we have to know about the tower, the earth, and our location on the earth to figure out the angle of the slant?"

A couple of weeks later, I had the pleasure of meeting another devotee of the St. Lawrence Seaway and river, the surveyor Brent Collett, whose office on King Street in Brockville sits within 200 metres of the river's banks. Not long ago, Brent renovated the capacious multi-suite building that houses his office and renamed it the Chipman Centre, after the distinguished 19th-century surveyor Willis Chipman, another votary of the river and perhaps even more so of its shores and towns and their inhabitants. Indeed, on page 70 of this book, in a sidebar on Chipman, readers will discover that the man's concern for the planet extended brilliantly, at times heroically, to the planet's freedom from pathogenic contamination and to the effects of that contamination on human beings both in Ontario and across the continent.

On the day I visited Brent, I also met his one-time mentor, the surveyor Ken Wiseman, who sold Brent his Brockville firm in 1989 and who, that afternoon, drove me to the old St. Peter's Cemetery just west of town. There, within throwing distance of the river, we gazed solemn-faced at Mr. Chipman's well-surveyed grave, which, like that of any human being thus interred, is his final and most integrated collaboration with the earth. A few days later, I was informed by a surveyor that, for earthly intimacy, the only relationship that surpasses the one with the grave is that of a man with his Scotch, the preferred sap of a certain generation of surveyors. What he said he loved most about the *auld drynk* was that it tasted of "the ground, the peat, the highland stream."

The question is: Do the ambitious young surveyors of the present possess even a remote version of the planetary connections of their mentors and forebears? I have been urged not to characterize Ontario surveyors in terms of who they were and what defined them 50 years ago. One veteran surveyor told me, "The kids these days are a new breed, fixated on technology, determined to work in the cities, detached from the fields and forests.... Ninety percent of

Four generations of Brockville surveyors: from bottom left, clockwise, Raymond Mucklestone, who began his career in Brockville in 1942; Martin Kaldeway, who articled to Mucklestone; Ken Wiseman who also articled to Mucklestone; and Brent Collett, who articled to Wiseman. In one way or another, all four surveyors connect back professionally to Brockville's most famous surveyor, Willis Chipman, whose records, some of them more than 130 years old, now reside with Brent Collett Surveying Ltd. on King Street.

the time they're dealing with the world through a computer screen or a GPS unit!"

It would be easy enough to embrace such a statement as empirical, except for one thing: It is not. My own observations tell me that while the young practitioners are indeed a new breed, with a fascination for technology, they continue to embody a version of their forebears' allegiance to the land. Of the half-dozen young surveyors whose lives and futures came into our story a few pages back, Simon Kasprzak, for example, is all but certainly a child of the spruce forests and the Ottawa River, where he was raised in his native Renfrew. Sophie-Rose Côté is a skier and sailor, and during her years at the University of New Brunswick was a member of the UNB Woodsmen Team, which travelled the country competing in events involving axes and two-person crosscut saws and floating logs – a hearkening not just to the classic lumberjack or "lumberjill" but to the old-fashioned line-cutting and bush-clearing of the Canadian surveyor. Not long ago, Mel Truchon addressed her appetite for the outdoors by purchasing 20 acres of forest, pasture and pond, near Huntsville, where she keeps a pair of horses and welcomes a regular procession of bears, moose, deer and other creatures.

As we have noted, Jansky Lau and Boney Cherian chose surveying at least partly because of their discovery of outdoor Ontario – and because of a collateral awareness that their ensuing careers would be spent if not exactly in the wilds of the north at least occasionally in the rainforests and mountain passes of Markham and Mississauga.

Meanwhile, there is another sustaining trait common to surveyors – and I do not mean a fetish for the log tables or for historic or digital theodolites (these too of course may factor soothingly into their lives). The trait begins, rather, with the persistent curiosity that so clearly shapes the thinking of the average surveyor, and finds its higher indices in that oldest of coding and preservation mechanisms: the story – the urge both to tell stories and to listen to them. Is there another profession, anywhere, that is required by law to assemble data on every application of its art and science, and to preserve that data as a reference, in perpetuity? Until a few years ago, surveyors in Ontario were required to keep notes corresponding to every

In order to dispel the isolation of long summers of line-cutting, survey crews in the early 1900s occasionally hauled Victrola gramophones into the Ontario wilderness, along with a few popular records such as Billy Golden's "Turkey in the Straw," John Philip Sousa's rendition of "A Hot Time in the Old Town Tonight," or any of the grand operatic recordings of Enrico Caruso (all available today on iTunes).

element of their field work: from chance meetings and wildlife encounters to crew disputes, bug infestations and astronomical events. (Well into the 1800s, the surveys themselves were considered so important to civil stability in Ontario that the penalty for tampering with a survey stake or monument was hanging – which is of course a bizarre story unto itself.)

The Aboriginal people of Australia are famous for their songlines – in effect musical stories that they have preserved and sung for generations as a way of remembering, say, their migration routes across the outback, the location of shelter and springs, as well as key elements of the moral and social codes that have sustained their culture across the centuries. The storytellers of ancient India memorized the histories of their cultures in oral poetry – histories that are still recited at lavish yearly festivals.

Roll your eyes if you wish, but there is more than a passing correlation between the traditions of the ancient storyteller and the obligation or desire of the surveyor to preserve his or her record and values: in notebooks, plans, maps, in stories told and heard – all of which, in their way, carry the code, the cadastre, the way forward. The notebooks from Ontario's Great Survey of 1901, perhaps the most ambitious and extensive survey in history, are a rustic masterwork, Odyssean in scope, on early wilderness travel and surveying in the province's north (What is Homer's *Odyssey* if not a kind of trail guide, a courage maker, for those coming after?). The notebooks touch upon forests, waters, minerals, metals, First Nations – abundance and hunger; wolves and waxwings; survival and mortality. And of course surveying on a magnificent scale. And on an immediate scale, too. The surveyor Hume Proudfoot's notes from the Great Survey include commentary on the qualities of cedar, hemlock and spruce in the making of boundary markers; and on how the hematite in the surrounding rock messed briefly with his compass reading at a particular bend of a remote creek; and on the importance of stopping for blueberries along the Albany River, at west longitude 88 degrees.

The Ottawa surveyor Dave Woodland told me about notes from the mid-1800s, written by an Ottawa surveyor who had gotten a contract for a job in a remote part of eastern Ontario, in mid-winter, had hired chainmen and axemen in a nearby village, and had led them out into the forest. "These rural jobs could be heaven or hell," Dave said. "On this one they got socked in by several days of heavy snow. They were living in tents, totally miserable, weren't getting any work done. And then they ran out of food. So the surveyor set off for the nearest village, maybe twenty miles away. When he left, he wrote something like, 'Snow falling. Men drunk'... and when he got back the next day, 'No break in weather. Men drunk.'"

The underlying social messages of the notebooks could be coarse and prejudiced. At the Surveyor General's office in Peterborough I saw notes from the early 1900s that included a faded photo of an aboriginal woman labouring under what

would seem to be a 30-kilo load of survey supplies, while a related photo showed a male member of the survey crew carrying a load of perhaps half that weight. The same notes refered to the female camp cook as "nothing to look at, with a face marred by smallpox" – later praising her for her "capability with bannock and rabbit stew."

Other notebooks bear surprise testimony to global events of their time. Surveyor Shirley King's notebook from an Ontario-Quebec boundary resurvey, in 1931, includes a rare photo of Charles Lindbergh's Lockheed Sirius descending onto the Moose River near King's camp. The celebrated aviator and his wife Anne were on what they called a "survey flight," by which they hoped to find the shortest practical air route from New York to Tokyo. When they put down to refuel in Moose Factory, they met King and his crew, and King snapped their photo.

Brent Collett related a tale about a diary written in 1876 by the surveyor Henry Lillie, who lived in the village of Lyn, just outside Brockville. Lillie's entry from July 27th recorded that he'd done his chores in the morning, had eaten a noon dinner in the village, and had set off for Brockville to do some surveying along the St. Lawrence River. "What had brought me to his notebooks," Brent explained, "was that the surveying he intended to do was in precisely the spot where I myself was going to be doing a survey. The twist was that Lillie took his kids along that day, because the circus had hit town. Off they all went to see it, and Lillie ends up describing in detail what he calls the 'wild people, the dwarfs,' as well as 'the monster man,' who he depicts as being 'seven feet eleven inches tall and proportionately large at 580 pounds.' I read all of this, quite fascinated, but at the same time wondering when he was going to mention the actual surveying. And finally he did, in one last little snippet, which fortunately gave me the information I needed."

On some level, surveyors know well the underlying spirit and themes of the tales they tell: patience, survival, uncertainty, the endless intricacy of humanity's collaboration with the planet; the planet's simultaneous resistance and embrace.

From the beginning of my work on this book, I asked surveyors for their stories, and got them – often brief, distortional, quirky: Gord Keatley's quaintly bucolic story (page 32) of surveying as a kid on Manitoulin Island with Ace the canine OLS.

And Charlie Fairhall's story (page 133) of having a spry old client drop dead on him, mid-winter, mid-job, practically mid-sentence.

And Mart Himma's story (page 48) of

Opposite: Anne and Charles Lindbergh's Lockheed Sirius floatplane landing at Moose Factory en route to Japan, near the camp of surveyor Shirley King, August 1, 1931.

the day an Ottawa survey crew inadvertently shut down a section of the capital and triggered the full deployment of an RCMP bomb squad.

On another day, I winced at Talson Rody's tale (page 10) of falling through the ice into a bath of concentrated beaver feces, and in the sub-zero wilderness having to work for two days in his dung-saturated clothes.

One day Blain Martin told me how as a student in the 1970s, he had come to the end of a long day of surveying in the graveyard at St. John's Anglican Church in York Mills, and had been left on his own at dusk to finish up the day's assignment. He said, "Finally, for a few minutes, I was free of authority." At least until he looked down. "I couldn't have been more spooked," he said. "There in front of me, entirely unanticipated, was the grave of Colonel A. J. van Nostrand, one of the founders of Speight, van Nostrand & Gibson, the company I was working for at the time."

At the meeting of the AOLS Eastern Regional Group in January 2015, Harland Moffatt recalled an old-time ethics-bender involving a Parry Sound contractor who at some point nearly a century ago was hired by a surveyor to cut line through several miles of wilderness so that a road could be surveyed and built. Harland said, "When the surveyor showed up a few days later, a huge tract of forest, three or four square miles, was simply gone, burned to the ground by the contractor. The surveyor explained to the guy that this probably wasn't the best thing for the wilderness, the wildlife, and so on – to which the guy responded, 'Ah, go to hell! This is the way it's done up here! Cutting a line with axes is a waste of time.'"

Henry Lillie

At a meeting of the AOLS Archival and Historical Committee, I heard the surveyor Peter Moreton describe how, as a teenager, on his first day with a survey crew, he was tricked into taking a powerful electrical shock when the distant end of a steel tape that he had been instructed to hold was placed intentionally on a high-voltage electric fence.

At the same meeting, the surveyor Gord Good told a story about a tripod, transit, compass, etc., that had been found at the bottom of Beverly Lake, west of Brockville. He said, "Of course this raised questions: Had a boat tipped? Had a crime been commited? Should the equipment be salvaged as evidence? Should it be cleaned up for somebody to use?" Vicky Culbert, the wife of the Goderich surveyor, Doug Culbert, speculated that a surveyor's wife had probably grown tired of her husband being "constantly away on surveying jobs," had rowed his equipment out onto the lake and introduced it to the depths.

A week later, Georgina Smith told me that when her husband Andrew was a young surveyor in the London area, his boss sent him out one morning to do some field work alongside another novice. In order to accomplish their day's task the pair had to use machetes to cut brush in a dense woods. "At a point when they were well into the trees," explained Georgina, "the fellow said to Andy quite solemnly: 'Do you ever feel like just whacking somebody's head off with one of these?' Andy said, 'Well, I can't say that I have.' And neither of them said an-

Tom Bunker • The Charting of a Life

Tom Bunker, the esteemed Gravenhurst surveyor, has survived field and personal battles that few in the profession ever have to face. In 1979, while managing the Timmins office of T.E. Rody Limited, he and a couple of assistants travelled to Moosonee to participate in a mapping project. "One of our survey stations was out on the James Bay flats," he said in September, 2015. "To get above the trees so that we could see what we were doing, we had to build a tower."

For that reason, when Tom and his crew boarded a helicopter on the morning of August 16 to travel to the flats, they were ferrying a heavy load of two-by-fours slung on a cable beneath the aircraft. "I was sitting up front beside the pilot," Tom said, "and as we moved out over the spruce forest, we heard a sharp bang from the rear of the helicopter. The pilot said, 'I think we just lost our tail rotor.' Apparently, the lumber had swung back and sheared it off. It was obvious right away what was about to happen, because the helicopter started corkscrewing. I shouted at the guys in the back to get their seatbelts on, that we were going down."

Fortunately, as the helicopter spun to earth from about 500 feet, the pilot was able to scream a mayday into the radio, so that when the crash came, those back at the airport in Moosonee knew where to find the wreckage, about half a kilkometre north of the Moose River.

Within minutes, a rescue team was on its way downriver by boat. From a point alongshore, the team hacked its way inland with machetes. "We went down in very dense spruce," said Tom. "They stretchered us out one at a time. I was the last one to go, because I didn't seem to be seriously injured."

It was not until Tom reached the hospital at Moose Factory that he realized his injuries were more severe than he had imagined. He and two others were flown immediately to Kingston for orthopedic care.

Looking back, Tom says he was lucky. "We all had broken backs. The pilot was left a paraplegic. I required spinal fusion of two crushed vertebrae. But after a year and a half I had pretty much regained my mobility."

While the significance of Tom's story lies at least partially in the drama itself and in the heroics of rescue and survival, it finds broader meaning in the way a momentary confluence of forces can rearrange the charting of a life.

Within a year of regaining his mobility, walking became painful for Tom. It was discovered that a spinal-treatment drug had cut off the blood supply to his left femur, causing part of the bone to collapse. "The only thing that would fix it was a hip replacement," said Tom. But at age 32, he was considered too young for the procedure.

Determined to move on, Tom returned to school at Laurentian University in Sudbury – first for a single course to improve his business skills; then full-time for a degree in commerce. He was already an engineer and of course a surveyor, and on weekends while at university travelled from Sudbury to Timmins to survey for his old firm. When the routine became a grind, he gave it up and, upon graduating, took an accounting job with Coopers and Lybrand. In 1987, he qualified as a chartered accountant. But he was a surveyor at heart, and during that same year he returned to the fold as a condo specialist and comptroller in the Whitby office of Horton, Wallace and Davies. Before long, he quit that job, too, and became a surveying consultant, hiring out to the AOLS. Finally, in 1989, ten years after his accident, he found the independence and opportunity he seemed to crave, by purchasing a surveying firm in the Muskoka town of Gravenhurst. During the decades that followed, he trained his brother, Chris, and son, Tim as surveyors – and in 2014 sold them the firm.

In early 2015, I asked Tom if he possessed any redeeming memories of the traumatic day that reshaped his life so many years ago. He said, "Apart from being grateful for everyone's survival, I remember very clearly crossing the river from the hospital in Moose Factory to the evacuation plane in Moosonee at four in the morning. I was on a stretcher laid across the gunnels of a freighter canoe. The sun was coming up. The water was lapping beneath me. I couldn't see it. It felt dangerous. It had been a long day." At this point in his story, Tom paused, smiled, and said quietly that he was hoping there wouldn't be any accidents.

Opposite page: Tom, with tellurometer, at Gull Lake survey camp, circa 1969.

Right: Tom (seated) with brother, Chris (left), and sons, Tim and Jason.

Below: Gavin Lawrence, OLS #1906, in the mixed forests of central Ontario. At right: Tham Shanmugarajah on the banks of the French River south of Sudbury.

other word for some time. The day was a bit tense."

Others told me stories with the proviso that under no circumstances could they be included in the book. One such tale featured a trio of surveying students during the mid-1980s, doing summer work on a highway bridge, one lane of which had been closed to traffic. The storyteller explained that about 200 metres from the bridge a traffic sign served notice to approaching motorists: "Surveyors at Work – Prepare to Stop." Another sign, further along, said, "Stop Here."

The gist of the story was that on a midnight sojourn from camp, equipped with a watercolour brush, an ounce or two of paint and a flashlight, the students altered one letter on each sign, so that in the morning motorists were greeted by the message: "Surveyors at Work – Prepare to Strip" ... and by the subsequent directive, "Strip Here."

On the same theme, the surveyor Gavin Lawrence described a day shortly after his arrival from South Africa in 2003: "I got a call from my manager asking me to drop everything and go to a house on the lakeshore in Oakville because they were putting in a pool and we needed to stake the property line. If a line is shared mutually with a neighbour we normally knock on the neighbour's door and let them know what we're doing. So I walked next door to a big fancy house – had on my safety vest, helmet, work boots. What seemed odd for the time of day was that there were quite a number of very high-end cars in the driveway and along the street. As I got close I could see through the windows that there was a party going on: liquor, decorations, well-dressed women laughing and celebrating. I rang the bell and was waiting when suddenly a woman flung the door open, stepped out,

grabbed my arm and screamed back into the house, 'Girls, the stripper is here!'"

Further yet to the theme, Susan MacGregor described being sent as a summer student to survey at a nudist camp. She said, "The thing is, when you're a surveyor you have to learn *not* to pay attention to certain things – you're often walking into people's backyards, invading their privacy, walking right into homes in some cases. At that point in my life I was fairly naïve – had some pretty wild misconceptions about what a nudist camp might be. In the end, what opened my eyes wasn't the nudists – they were just people, *very* ordinary, living as they chose to live. Rather, it was a sense of how we nurture our misconceptions and unfortunately don't always *get* to the nudist camp, or wherever reality lies, to get them corrected."

Susan tells another story about an expropriation in which she was involved during her years at the Ministry of Transportation, when Highway 416 was being built in eastern Ontario. "Unfortunately, we had to take land from an elderly farmer – not for actual road-building but because a number of communications towers had to be moved, and this man's farm was the only place to put them. He'd known about this for some time, had fought every possible fight in order to keep his land, and had lost them all. And now I was at his door,

the final blow, to let him know that I needed to walk the property to make sure the survey markers were in place so that we could file the expropriation plan. This last step would put his land in the hands of the Crown. I'll never forget it; he was such a gentleman. Rather than curse or threaten me or complain, he decided he'd walk the property with me. We set out, and after a few minutes he began telling me stories about what had happened on his land over the years – which of course is one of the things that interests us as surveyors: what happens on land. He said his dad had walked this same route with him when he was a boy. It was heartbreaking; I'd gone there thinking about what *I* was going to say, about how *I* was going to handle myself, and the lessons that came out of it lay entirely in what *he* had to say, in how *he* handled himself – his strength and respect."

Along the way, I talked to surveyors who had *lived* their stories – sometimes with significant courage and resolve. I am thinking, for example, of the persistence and survival of the surveyor Tham Shanmugarajah, who arrived in Toronto from Sri Lanka in 1984, with a degree in architectural science, newly married, penniless – "shocked and confused by this new culture and environment," he says.

Unable to find work in his field, Tham took a low-paying job in a Toronto factory, living for months with his wife in a crowded basement apartment. "One day," he says, "I heard about a night-school course in surveying skills, being held in a local elementary school." He took the course, finished it, and got work as a chainman for a few months. Dreams of bigger things led him to Ryerson, where he studied geomatics and surveying, eventually graduated and articled, and was commissioned as a surveyor in 2001.

The following year, fed up with working for others, he started a business of his own, still broke. He said, "It was me and three other guys, two of whom had to quit, because there was no money to pay them. I was living on my line of credit, working day and night, raising kids, barely able to buy groceries. I spent a lot of hours driving around in a little used car, trying to find jobs." Tham recalls the Vietnamese draftsman, Michael Nguyen, a graduate of Seneca College, showing up at his door one day in 2003. "He had a stutter; nobody would hire him; he begged me for work, promised he'd prove himself. I thought, *wait a second, I haven't even proved myself.* But I took him on – taught him everything I knew about drafting from my training as an architect. I didn't pay him much, because I still had no money. At this point, I'd been struggling for nineteen years."

Today, Tham Surveying Limited, operates successfully in Vaughan. Tham's Vietnamese draftsman ("one of the best there is, indispensible to the business") is still with him. Tham's two sons have fin-

Tham Shanmugarajah with his trusted draftsman, Michael Nguyen (left), hired in 2003, when according to Tham there was little money to pay him.

ished degrees in geomatics, at Ryerson and UNB, and one has begun articling as a surveyor. "They'll have their own struggles," says Tham. "Fortunately, I've cut some brush for them. Mine was a different kind of struggle."

Michael Griffiths faced his own challenges on arriving from Jamaica in 1980, in possession of a university land surveying degree and surveying experience, and keen to qualify as a professional surveyor. "The problem," he says, "was that my degree wasn't accepted here; it meant nothing. So I took surveying work at a lower level." After some 13 years of resisting what he knew he had to do if he wanted to gain his commission in Ontario (basically, start from scratch at an Ontario university), he gave in, started over – went to Ryerson for three years, to the University of Toronto for two, and then did four years of articling. Only then, 22 years after arriving in Canada, was he able to place the letters OLS after his name – and eventually assume a role as a key member of the Surveyor General's office in Peterborough.

Lorraine Setterington arrived at the gates of the profession not from another country but from a sequestered realm that, prior to 1969, when she earned her commission, was as distant from the offices of Ontario surveying as were the physical worlds of Sri Lanka and Jamaica. In May 2015, more than two decades after her retirement, she said, "It never even *occurred* to people in the 1960s that a woman could be a surveyor. My daughters don't understand that when you got married in the 1950s, as I did, you had no expectation of working outside the home. I had my kids, I sewed their clothes, knitted their sweaters, gardened all summer, canned in the fall, cooked and cleaned all winter. When my husband, Bill Setterington, became the chief engineer in Leamington he made an agreement with the town that allowed him to use the equipment on weekends to do surveys. Like many wives, I went out and held the other end of the tape. I remember the day we were driving down Talbot Street and he said, 'You know if you're going to keep doing this, we should see about getting you a licence.' I had no clue at that time that there weren't any women in surveying. But I went ahead, got my trigonometry at the local high school, got the articles ready, and Bill and I took them into Toronto to the Association office, at the corner of King and Yonge. John Bradbury, who was in charge, looked through the paperwork – everything was okay. In those days, they just signed your articles on the spot. Bill made the mistake of saying, 'You'll notice this is for Lorraine.' Suddenly, John got this asphyxiated look on his face. He said, 'I can't sign them for a woman.' And he didn't! Either he didn't believe women should be surveyors, or didn't feel he had the authority to proceed. Or both. That was in July.

"Apparently, in November John got a call from a reporter at the Toronto *Telegram*, who wanted to do a feature on me – presumably about my difficulty getting into surveying (somebody had tipped them off). I was nine months pregnant by this time, and pretty much in labour, so I couldn't see the reporter, and the story never got written. But for some mysterious reason my articles were approved immediately."

A year later, when Lorraine was ready to try her first set of exams, the attitude of the Association had not changed. "But mine had," said Lorraine. "In those days we wrote course exams in a building on Bay Street opposite where the Sutton Place Hotel used to be. At the door of the examination room the folks in

Lorraine Setterington, the new face of surveying in Canada, photographed near Leamington in 1969.

charge insisted I was in the wrong place – told me *these* exams were for surveyors. 'Yes, I know,' I said, very courteously, and I walked into the room and sat down."

Six years after that, the president of the Association, John Barber, phoned Lorraine and asked her if she'd be interested in filling in for a few months as Association secretary, the organization's top administrative job, while Barber and the executive cast about to fill the newly invented role of executive director. "The previous secretary had quit in a huff," said Lorraine. "So I started in. The office was at 6070 Yonge Street, between Finch and Steeles."

Lorraine's first heavy challenge was to track down 75 young aspirants who had at one time or another filed articles to become surveyors but who had not been heard from in years.

She said, "I took all seventy-five files home one night, sat down at the dining table and, over a period of several days, wrote a personal letter to every one of those guys. And gradually got their responses. Turned out half of them had left surveying, had given up. One who hadn't, Rick Dickson – a guy who eventually became president of the Association – explained to me that he and the others had been writing to the office all along, wondering about their status, asking when they could try their exams. He said, 'The Association never wrote back – never did anything!' I said to him, 'Well, that is about to change.'"

When six months later Lorraine asked for the substantial raise in pay that had been promised her if her employment worked out, she was granted it without even an interview. She said, "At that point, they decided there was no need to hire an executive director, that I could handle that job too. So they offered it, and I took it. I had two women working for me. Otherwise, I was director, standards officer, secretary, treasurer, public relations officer, disciplinarian, shoulder-to-cry-on, anything you can think of. Me and six hundred men."

Drew Annable, another of the seventy-five whose entreaties to the Association had gone unanswered prior to Lorraine's arrival (and who, like Dickson, eventually became president of the Association) says, "I owe a lot to Lorraine. We all do. She sorted things out."

Meanwhile, Maureen Mountjoy, the second woman commissioned in Ontario, in 1978, faced challenges every bit as prickly as those faced by Lorraine. "Oh, the early years were nasty," she said in an interview in August 2015. "I articled at Red Petzold's office, and Red himself was always very supportive, always a gentleman. Lots of times, he and I would go out on a job together, just the two of us. I would run the instrument and he would be the party chief. I learned a lot

Left to right: Maureen Mountjoy, Anne Cole and Kathy Sam-Guindon, Ontario's second, third and fourth female surveyors.

that way. In those days, you made decisions in the field. You came across monuments, you surveyed, you decided if the monuments were good or not, and you made your assessments on the spot. By the time you returned to the office, you knew what your boundaries were. Nowadays, they gather all the information with total stations and GPS, analyze it back in the office, and make boundary decisions at their desks."

It was in Red's absence, while Maureen was working with the survey crews, that her apprenticeship tended to go awry. "Some of the guys would say terribly demeaning things to me out in the field, thinking I'd just quit. I'd go home and shed a few tears and tell myself, I am *not* going to let them make me quit! The next day, I'd be back in the office, pretending it didn't bother me. The ironic thing was that eventually I became their boss, and that didn't sit too well either. One fellow actually left the firm and moved away because he couldn't stand having a woman in charge. A few years later he wrote to me and said, 'I apologize, Maureen – I was an old fool.' I still have the letter – I sometimes take it out. More and more as the years pass, I think, *I am so glad I persisted…* If I hadn't I would have missed all these wonderful years working in a great profession."

Not long ago, Maureen mentioned a striking footnote to her apprenticeship: "Because Red Petzold was in and out of the Association office so often, consulting with Lorraine about my articles, those two developed a romance and eventually got married. They both used to joke with me that I was to blame."

THE LARGER STORY OF WOMEN in surveying in Ontario is still being told – a narrative stretching from a complete

absence of women in 1968 (76 years after the Association's founding) to one woman in 1969, two in 1978, four in 1982, and today nearly 40, some seven percent of the surveyors active in the province. In 2009, Susan McGregor became the first female Surveyor General of Ontario, and in 2014, Dasha Page became the first female president of the AOLS. Young female surveyors such as Sophie-Rose Côté, Vicky Kumar and Mel Truchon have taken their place among the province's most progressive companies, and the Association is a dedicated advocate for an improved gender balance in the profession. In 2015, the AOLS began offering awards to female geomatics students at York University, and it

Crystal Cranch • What I Considered Freedom

"I **was sixteen years of age when reality blindsided me.** I was sitting in my Grade 11 Biology class. My teacher, Ms. Seigel, was yelling at me, although I was oblivious to it as I gazed out the window at what I considered freedom (already I knew that happiness for me was connected to the great out-of-doors). Later, I asked Ms. Seigel if she liked her job. With one exasperated look, she gave me her answer, and I understood that I would not become another sad statistic in a job satisfaction survey. Eventually, I found my connection to the outdoors in the profession of surveying. It has never let me down. Even today, when the phone is insistent and the demands are overwhelming, a few hours in the field invigorate me, take me back to what it was I originally sought in the profession."

"I consider myself a supplier of knowledge, not a supplier of plans."

You Know You're a Surveyor When:

- *your friends and family can identify the location of several thousand survey bars because you seem to be incapable of passing one without sharing your knowledge*
- *you know the legal description of your property, but cannot quite remember the street address*
- *you're walking down a gorgeous sand beach and you do not notice the feeling of warmth on your feet, or hear the sound of the waves, because you are too busy trying to figure out who owns the beach*
- *you understand how we can use satellites to determine our precise location on the planet, but you still cannot figure out the two-line phone*
- *someone asks you how far to the nearest pub, and you give them the answer to the nearest millimetre.*

Mart Himma • The Bomb Kettle and the Liquid Bite

"**When I was articling back in the early 1980s,** my wife, Line Delorme, was articling with Fairhall and Moffatt in Ottawa. At some point in I think 1982, there was a terrorist scare in the city, so the whole place was in a heightened state of alert. One day shortly after the scare, when Line was at work, the firm got a call asking if they'd send an employee to a particular address to identify what was probably a survey instrument, but might be a bomb. So my wife's boss asked her if she'd drive downtown to identify this equipment. He said, 'We've got a crew doing a survey there, so we should see what's going on.'

"As Line headed into the area, she came to a barricade of RCMP officers quite a ways from where she wanted to get to. She explained who she was and that she'd been sent by the firm, and one of the cops got on his walkie-talkie and said to somebody further along that she was coming through – and a minute later she arrived at a much bigger barricade, where they gave her a pair of binoculars and ask her to look down the street and identify what was on the sidewalk.

"As she's looking through the binoculars, she sees a guy in a bomb suit – full armour. They've got the bomb kettle, which is a little trailer that they put bombs in to explode them. He's got a long stick and is about to lift up the object by the leather handle on its casing, which resembles a kind of steel bell jar with latches.

"What none of them knew at that point was that a half hour earlier the survey crew from Fairhall amd Moffatt had packed up and had left to go for a burger and maybe a bite or two of liquid lunch. But when the crew packed their car, they mistakenly left this piece of equipment, with its casing, sitting on the sidewalk. And of course when they tried to get back to the site an hour later, they couldn't because of the barricades, and didn't have a clue what was going on. So they simply took off for another 'bite,' figuring maybe the traffic would clear up while they were gone.

"Meanwhile, Line looked down the street with the binoculars and said, 'Yes, that's a T1A case, and it does look like one of ours.' And the officer gets on the walkie-talkie again and talks to the bomb guy, who is just about to drop the T1A into the kettle to blow it to smithereens. Very cautiously, he puts it on the curb. The thing in its case is shaped like a diver's helmet or a bell jar. At that point the bomb guy notices the two clips that secure the case and undoes them. He then ties a string to the leather handle on top and yanks on it, knocking the thing over as he pulls the lid off. The cops give the binoculars back to Line, and she says, 'Yes, that's definitely a survey instrument – it's ours.' So they pick it up, and a minute later they hand her the T1A in its case, and she takes it back to the office and puts it on Charlie Fairhall's desk.

"Eventually, the crew returns. They go into Charlie's office and start explaining what a terrible day it's been, how there must have been some terrorist activity or something, because they couldn't get back to the site after lunch; it was all blocked off by the police... they didn't get much done, etc., etc.

"Charlie had quite a sense of humour. He hauls the T1A up onto his desk and says, 'Fellas, do you recognize this?' Which of course silences them, because of course they recognize it. He says, 'Well, the next time you leave it on the sidewalk somewhere, could you perhaps put a sign on it explaining that it's not a bomb? That way we could get a day's work done, and get it done without having to tie up half of Ottawa and a dozen members of the RCMP for an afternoon. We're surveyors, not terrorists.'"

Erindale grads Line Delorme, OLS #1630, and Mart Himma, OLS #1631, earned their commissions as Ontario Land Surveyors in 1988.

Below: In February 2015, at the AGM in Huntsville, Dasha Page became the first female president in the 125-year history of the AOLS.

supports the ambition of York's Lassonde School of Engineering to be the first of its kind in Canada to achieve a 50-50 gender split.

THE AMERICAN MYTHOLOGIST Joseph Campbell said the history of anything resembles a field of endless dots to be connected, and that any one of those dots could be chosen as a beginning point. If you took a LiDAR scan of the history of surveying in Ontario, it would give you a billion, perhaps a trillion, such dots –

and an infinite variety of ways in which they could be joined. We have chosen to begin our tale with the dot that is Gordon Gracie's introduction to the Erindale Survey Science program (and will fill in more dots from the Erindale years in an upcoming chapter). We could as easily have begun with the dot that marked the arrival in Upper Canada of Ontario's original surveyor and mapmaker, Samuel de Champlain, some 400 years ago. Or the one that marked the formation of the AOLS in 1892. Or any of numerous dots that marked, say, The Great Survey of northern Ontario in 1900, or the boom years that followed World War II and marked the beginnings of modern surveying in Ontario, or the dots that marked the technological revolution of the 1980s and '90s.

I might note in passing that my own small dot met the telling of the story on a scorching day in August 2014, at Pearson Airport, within a few hundred metres of where Gordon Gracie landed on his flight from Washington in 1971. As I left the terminal to fetch my car, I got a call from a publishing friend, Barry Penhale, telling me that the AOLS was planning a book, a history, to celebrate its 125th anniversary, and that if I was interested in being involved I should call Maureen Mountjoy at the AOLS office. I did so within minutes and perhaps six weeks later found myself in the board room at the AOLS offices, chatting amiably with the task force that had been established to bring the book into being.

It would be convenient to declare that the rest is history – in book form. Except that it is not. Because, from the beginning, the task force and I were determined that the book should not become a formal schematic history, or any mere chronology of events. The idea rather was that it should be a celebration of the years of surveying in the province – a kind of free-form birthday party for the roughly 2000 surveyors who have operated in the province since the mid-1800s, and for the Association that binds them as a profession.

Above all, we wanted the book to be personal – a record of people, their stories, their voices, their images; their connection to the planet and what it is they do on it. As for the early years of the AOLS, we were determined that they, too, should be conveyed as a story – one of people and places, and of the struggles and accomplishments that gave shape to the profession in its infancy.

My earliest research led me *deep* into the AOLS archives – into books and documents, some dating back to the early 1800s; into photo and map files and notebooks. Along the way, I read dozens of old issues of the *Ontario Professional Surveyor*, the quarterly magazine of the AOLS, as well as John Ladell's well-loaded but largely ignored history of Ontario surveying to 1992, titled *They Left Their Mark*.

Perhaps most enjoyably, in those early days, my freewheeling research put me on the highways of Ontario: to Windsor, to Ottawa, to London, Cornwall and Northwestern Ontario.

One of the highlights of those road

At left: The marble pillar that marks the grave of David Thompson in Mount Royal Cemetery in Montreal. Centre: A drawing said to be of David Thompson during his wilderness years, but most likely done many years later. Right: A highway boundary marker, carved from Ontario limestone during the 1920s, and placed, along with thousands of others in the province, to mark the edges of road allowance along Ontario's 66-foot-wide highways.

trips came in late May 2015, when I drove from Cornwall into Montreal and fulfilled a longstanding intention to visit the grave of David Thompson, history's greatest surveyor and geographer. Thompson's remains lay unmarked in Mount Royal Cemetery for some 70 years after his death. It was not until 1926 that a modest marble column was erected on the grave. Unfortunately, a bronze casting of Thompson's sextant, which once sat atop the column, was stolen twice over the years, with the result that its replacement is kept in a nearby tool shed, leaving a lead post, an ugly thing, protruding from the top of the column. The relative obscurity of Thompson's life and death is reflected today in our utter lack of awareness of what he looked like (for all his accomplishments, he never sat for a decent portrait). As of this writing, there is just one drawing of him from his wilderness years – an apocryphal likeness that does not show his face – and another from his old age, done in the jaunty-old-wingnut style, and surely no more authentic than the early drawing. On one historic website, a splendid and detailed image purporting to be that of Thompson is actually that of Alexander Mackenzie, Thompson's fellow explorer in the Northwest Company of fur traders.

At Kingston I visited the stone cairn marking the place where in 1783 the surveyor John Collins laid out the first township in Upper Canada, having purchased a quarter-million acres from the Mississauga tribe for a price that these days would fetch a miserly single plot in the average Kingston cemetery.

A few days later, in Windsor, I stood on the banks of the Detroit River, where, in 1749, French surveyors laid out the tiny seigniorial fields that marked the first European farming community in what is now Ontario. Seen today, *la Petite Côte* (the little coast), as the area was called, is a somewhat doleful patch of muck, whose boundaries are indistinguishable from the overriding English survey grid that knocked it off the map. Not surprisingly, by 1790, the area had become known as *la Côte de Misere* because its sandy soils had entirely lost their fertility.

From Windsor, my travels took me to Goderich, where I attended Doug and Vicky Culberts' annual barbecue for the Association's free-spirited archival committee. One of the features of the Culberts' property is a meticulously developed monument garden, a mini-Stonehenge, with fountain, that brings together old limestone survey markers from various Ontario highways, as well as an assortment of wooden corner posts and blazes, each suggestive of some 200 years of complex interaction between surveyors and the land.

The Perth surveyor, John Goltz, possesses no such physical relics. But in describing to me how as a rookie surveyor a half century ago he became enamored of the sun and stars, he provided me my

Doug Culbert surveys the monument garden that he and his wife Vicky have created beside their home in Goderich.

first glimpse of a dynamic that on the right day, in the right place, can still put a decidedly mystic edge around the art and science of surveying.

There is no obvious mysticism at the Surveyor General's office in Peterborough, but there is a clarity about the realities of history that resides not just in the framed relics on the walls or in the climate-controlled "plan room," where the maps and survey plans of two centuries are stored, but in the shapely and empathetic principles of the staff. During three days in the office, in March 2015, I was impressed equally by the stirring old notebooks and documents, by the loving attention afforded those documents by the likes of Eric Ansell, Donna Gertridge and Lisa Casselman, and by the openness of surveyors Michael Griffiths and Wikar Bhatti, the latter of whom spoke forthrightly about the racism he had encountered from one source or another during his early years as a surveyor. I was impressed, too, by the eloquent modesty of the Surveyor General, Susan MacGregor, who appears several times on these pages and who when I located her for an interview during my second day on the site was, as I recall, finishing a homemade lunch at her desk. True to form, she spoke not of her persistent travel and speaking schedule or the proliferating extent of her administrative and legal responsibilities (an extent made obvious by the foot-high piles of work folders awaiting her attention) but of the more personal elements of her job – how, for example, with kids, she had juggled home and career over the years, and of the necessity of remaining calm and afloat when the waters rose (as they apparently sometimes do) around her standard-issue MNR desk and office chair.

My trip to the Association's annual general meeting at Deerhurst Resort in late February 2015, was another lively round, bringing me into animated conversation with troupers such as Gord Good, Dave Woodland, Kathryn Sam-Guindon, Brian Maloney, Curry Bishop, Dasha Page, Anne Cole, Vladimir Krcmar, and many others. On a morning when the temperature outside the hotel had fallen to minus 40 degrees, and even the normally frost-resistant surveyors were stiffer than Giacometti sculptures, Steve Gossling told me about one of the earliest days of his inaugural summer job in surveying in 1971. It is a tale that in one form or another is endemic to a generation of Ontario surveyors.

He said, "We were doing a control survey along the Trans-Canada Highway between Wawa and White River.

"As the sun set that day, I was transported in a helicopter some twenty miles into the wilderness and dropped off alone at the first order station." His job, as he describes it, was to stay out there through the night, waiting for a signal that would be flashed to him along a cut-line across the hilltops. "When the signal came, I was to redirect my own light along a second line that had been cleared through the forest, so that the crew at one of the other stations could use it to complete their observations."

However, by 3 a.m., Steve had not received the signal. "Finally," he said, "I just trimmed my light and sat down by my fire, hoping a stray bear wouldn't find me. I guess I fell asleep, because the next thing I knew, the helicopter was landing, and the pilot was yelling, 'Jump in! We're low on fuel! We gotta go!'"

The Three Faces of St(Eve)... From the bottom: Sudbury surveyor Steve Gossling as one might encounter him today; Steve the novice surveyor, as portrayed in the accompanying story; and (in a foreshadowing of his future as an OLS) Toddler Steve, contemplating a model of the planet and the geophysical character of Niven's Meridian.

The reason there had been no signal, Steve explained, was that the line across the hills had not been cut wide enough. "So the entire night's work was wasted. Which for me, I guess, was a bigger lesson than anything I learned about actual surveying that day."

Steve's description of that lesson went as follows: "We are an imperfect profession."

That same afternoon, I chatted with Joe Young, the affable and energetic president of J. D. Barnes Limited, a company with some 375 employees (including more than 50 surveyors and 16 articling students); a dozen offices across the province (plus affiliates in Alberta, Quebec, and parts of the U.S.); and seven airplanes that the company uses in varied endeavors throughout North America and the Caribbean.

When I asked Joe what had enabled the firm (founded in 1960) to get so big and to remain prosperous for more than half a century, he responded without hesitation that "the magic" had started with the company founder, the late Jack Barnes, a transformative figure not just within the industry but in his support of the education of surveyors and in particular the survey program at Erindale College.

Under Barnes, Joe said, – as well as subsequent presidents John Barber and Gary Kirstine – "the company was constantly looking for and investing in new technology." It was, for example, the first firm in Canada to use electronic total stations; the first to have what Joe called "a big electronic plotter;" the first to have a computer division and to develop surveying software for the industry; the first to incorporate a mapping group, with "state-of-the-art mapping technology."

As Joe describes it, "Jack insisted on hiring and training good people and on listening to their views."

Joe attributes his own rise in the company to what his father taught him back on the farm, near Sarnia: "that hard work and dedication matter." His success has been no less a function of the support of those around him and of his willingness to keep learning. He pointed out that he had started at J.D. Barnes as a field assistant doing cadastral surveys and had eventually moved into the geodetic and engineering division. He said, "In 2002, I suggested to the company's executives that they open an Ottawa office," which

Clarence Monteith (left) and Colonel Sam Archibald met during World War I, when neither was yet a surveyor, and remained close friends throughout their long careers. Clarence's son, Jack, eventually articled to Colonel Archibald during the 1950s. Clarence and the Colonel are seen here outside the Colonel's apartment in London, Ontario, circa 1925.

they did and which Joe immediately volunteered to run. He said, "When our aerial mapping group had some problems in 2008, I volunteered to move over there and manage that part of the company."

Joe concluded his commentary by pointing out that, at the moment, there are just a handful of surveyors in the province under the age of thirty; that "you can make as much money in surveying" as in any other profession; and that, right now, surveying is "an incredible opportunity for any young person who is mathematically inclined or into sciences or the outdoors."

THROUGHOUT MY INTERVIEWING and research, the same names kept surfacing, from both the past and present: Fred Pearce (who, like Jack Barnes, was a committed proponent of progressive education for surveyors and was instrumental in the founding of the Erindale program), Oscar Marshall (another Erindale founder), John G. and John W. Pierce (famous for their work on the Manitoba-Ontario boundary survey), Jack Young (an Erindale professor and former AOLS president), James Ferguson (the GPS and LiDAR wizard), Brian Maloney (the former Surveyor General of Ontario), Harold Macklin and Patrick Monaghan (who with Oscar Marshall founded the famous Toronto firm, Marshall Macklin Monaghan), plus a parade of past presidents that included, among others, Wayne and Dave Brubacher, "Red" Petzold, Jim Nicholson, Howard Graham, Gren Rogers, Michael O'Sullivan, John Barber, Paul Church, Drew Annable, and the legendary Canadian Colonel, Sam Archibald. Sam was a veteran of both world wars and a mentor to the likes of Dane Wandabense and to the Sarnia surveyor, Jack Monteith.

On days when I didn't go to the surveyors, they came to me – generally to the library at the AOLS offices, where I had a desk with a nameplate that I had received (along with a one-hundred-dollar Tim Hortons card) for delivering a brief speech at the Veterans' Dinner at the Association's 2015 AGM. It was in the library that I had the pleasure of interviewing Izaak de Rijcke, Jansky Lau, Tim Hartley, Al Worobec, Bill Buck, Maureen Mountjoy, and others.

One day the retired Toronto surveyor and historian, Bill Daniels, came in and sat down. He is a gentle and erudite lone wolf who committed much of his life to the study and collection of antique surveying equipment. Bill can tell you anything you want to know about 200-year-old theodolites, pocket transits and circumferentors, or about the shops in Toronto that sold such equipment as far back as the early 1800s. In his prime, Bill travelled the province, dressing up in 19th-century costume, doing period surveying demonstrations at historical

Jack Barnes • Two or Three Stairs at a Time

Among 20th-century trailblazers in the surveying profession, nobody broke more trail, and broke it more confidently or effectively, than John Duncan Barnes – "Jack" as he was known throughout the profession. "He could see the future of surveying better than anybody I ever met," says his long-time friend and business associate, John Barber. "From the beginning, he knew what kind of company he wanted – and it wasn't going to be tied to the old ways of doing things."

Barnes also knew that in pushing forward, in expanding the idea of what a surveying firm could be, he was, as Barber puts it, "creating a kind of model that others could follow if they chose to."

Founded in 1960 on Bloor Street in Toronto, J.D. Barnes Limited was, by the mid-1960s, one of the biggest firms in the province, with its staff of 45, including six Ontario Land Surveyors. In 1965, the firm became the first in Canada to create its own geodetics division and, in 1972, the first to establish an aerial mapping detachment. During the early 1980s, the company purchased a computer programming enterprise and became the first to develop broad-application surveying software, which it sold both to governments and to other surveyors. Along the way, the firm was the first in Ontario to register a condominium plan (1967); the first in Canada to invest in what the company's current president Joe Young calls "a big electronic plotter" (1975); and the first to invest in an electronic total station, a Wild TC1, at a cost of more than $100,000 (1981).

"But it wasn't just technology that inspired Jack," says Young. "He believed strongly in hiring good staff and in letting them go creatively to work. I never worked with him; I joined the company about a year after he died. But I know from everything I've heard that his instincts in hiring were pretty much infallible."

Barber is candid in pointing out that every Barnes initiative didn't move smoothly into a profitable or productive future. "Sometimes we struggled with an innovation," he says. "You can't constantly be taking risks and not accept the possibility of failure. Where Jack differed from some executives was that he believed even the failures were a benefit both to the company and possibly to the industry."

Born in Toronto in 1927, Barnes graduated from Jarvis Collegiate in 1946 and articled with the firm of W.S. Gibson and Sons, before being commissioned in 1950. As a prelude to founding J.D. Barnes and Associates, he purchased the records of the firm King, Coons, Phelan and Porter.

Barber says that "as with all complex people," there were ironies in Jack's life. "He loved doing business, loved the challenges and risks. His sense of almost everything was: Let's get on this; let's get it done. Even physically, he was always in a hurry; he climbed stairs two or three at a time. And yet there was never any irritation attached to any of this, no pettiness or intolerance. He was fun to be around, made time for everybody. At a party, he was invariably the guy who was surrounded by people, all of them laughing."

Further to the theme, Barnes emerged into surveying at a time when no advanced education was required. "It was pretty much seat-of-the-pants," says Barber. Yet he was one of the foremost supporters of the academic survey program at Ryerson University during the 1970s, and was a committed contributor to the founding of the survey sciences program at Erindale in 1972. In 1973, he and the company committed $5,000 (plus $2,000 in each of the five succeeding years) to the newly founded AOLS Educational Foundation – a fund that now controls nearly half a million dollars.

Barber says, "Another interesting quirk of Jack's life was that he didn't marry until he was well into his fifties" – at which point he married Odette Lepage, a Quebec surveyor and the country's first female Canada Lands Surveyor (their son, John Jr., was born in 1986, just a year before Jack's untimely death). "And yet all through his longish bachelorhood," says Barber, "he was very much a family man, a kind of father figure, involved closely with his nieces and nephews – in fact with all sorts of young people, in school or at the firm."

According to Barber (who was president of J.D. Barnes Limited for fifteen years beginning in 1985), the lasting proof of Jack's genius lies at least partly in the growth of the company he founded to 375 employees today, and in the company's continued sense of innovation. "What made him such a pleasure and inspiration personally," says Barber, "was that he wasn't about ego. In his mind, you were as important a part of the company as he was; he wanted the best for the firm, not just for himself. Beyond that, he was always open to what you had to say, always excited about some new possibility or idea."

Above: Jack at play with Erindale Survey Science professor, Bob Gunn (second from left), and students Paul Miller and Ian Robinson (right), at Erindale's survey games in 1975.

Left: Jack (6′4″) as a newly commissioned OLS in 1953, with a quartet of his surveying pals.

Bill Daniels, in character as Surveyor General John Collins of Lower Canada, reenacts the creation of the first township in what is now Ontario, at Kingston, in 1783. Land for these earliest townships – created to accommodate United Empire Loyalists fleeing from the newly independent U.S.A. – was acquired from the Mississauga First Nation for a price said to be in the range of a few pennies per acre.

events and celebrations. Eventually he sold all but a couple of his prized and valuable antiques. When I asked why, he responded ruefully that for a variety of reasons he had had to quit surveying a number of years back and that (like artful eccentrics since time began) had at that point "needed the money."

On another day, the East York surveyor, Jim Hill, settled his rangy anatomy on one of two folding chairs that have been central to my in-house hospitality for the past six months. Jim has been retired for 21 years but was dressed for at least light duty in a plaid short-sleeved shirt with ballpoint pens in the breast pocket, and cotton work pants held up by two-inch-wide suspenders. He is a folksy, laconic, soft-spoken man who gives the impression of being perpetually amused and who has not missed a convention of Ontario Land Surveyors since Methuselah was a chainman. Jim spent his entire career with a variety of government ministries, including MTO at a time when the building of expressways in the province smoked more public dollars than did education or health care, which are by far the lead budget items today. "It was all highways back then," he said, citing a passel of those built during his salad days, including 400, 401 and 403, plus "dozens of old gravel roads that we were paving and turning into secondary highways."

In putting images to the changes that took place during his prime, Jim noted that the first chainsaw he laid eyes on in the 1950s – this after several years of cutting line with an axe and machete – "weighed a few hundred pounds and was said to be a convenience." He said that during his years at the Ontario Department of Public Works the first calculator he bought for the surveyors and technicians in his charge cost $600 (the equivalent of more than $5,000 in today's dollars) and was barely more convenient than a two-dollar slide rule.

Jim stated flatly that he has no interest in the esoteric side of surveying, viewing the profession rather as a practical endeavor aimed entirely at bringing, say, roads and bridges into existence, or creating and maintaining boundaries. When I asked if he ever looked back on his career in the context of the broader post-war culture – the consumption of resources and land and goods on a relentless new scale – he looked at me as if reassessing my intelligence and said, "No."

Jim Hill, seen here during the 1980s, was referred to by one of his peers as "the surveyor's friend in government."

Asked if there were stories he could tell me, he said he pretty much thought surveying *was* the story.

It was others who told me that during Jim's years of tendering survey contracts on behalf of Public Works – his years of staunchly upholding the statutes – he was "a bit of a teddy bear," never less than faithful in providing empathy and support to surveyors who he knew needed a boost. When I broached the subject with him, he said, "Even if it were true, I wouldn't admit it." Which, as a storyteller, is part of what I enjoy about Jim – his sense that what you withhold in telling your story is (as Hemingway reminded us) as important as what you include. "I have a few secrets," he confided, adding that one of his problems at his current age is that he "can't always remember what they are."

In the context of the AOLS, I have a secret or two of my own, one of them being – if it is not obvious – that I do not afford all of our early surveyors the level of esteem granted them in, say, the Archives of Ontario, or in a number of laudatory history books (it is a view shared increasingly, I believe, by the contemporary surveyors of the province). The reason is that, despite their talents and durability, more than a few of the earliest practitioners were hard-headed agents of the oftentimes ruthless colonialism that during the 18th and 19th centuries swallowed up most of what for ten thousand years had been the exclusive domain of the First Nations. Granted, it was a different time, a different culture, and people who might today have behaved more considerately acted badly in the name of empire- or nation-building. Mercifully, there were also early surveyors who did not represent the worst injustices of the period. During the 18th and 19th centuries, it was common for agents of European expansion – fur traders, explorers, surveyors – to take "bush wives" during their time in the wilderness. It was just as common for those agents eventually to abandon their wives, along with any number of their mixed-blood children. One who didn't was the surveyor David Thompson, who remained faithful to his First Nations commitments throughout his life: to his Cree wife, Charlotte Small, and his thirteen mixed-race children, most of whom accompanied him and Charlotte back to Ontario when his wilderness days were over. The late-18th-century surveyor Augustus Jones was another brilliant practitioner who maintained close and sympathetic ties with the First Nations – ties that, given the standards of the day, were viewed by some as scandalous, by others as politically and socially progressive. Whatever the view, Jones's links to the great chief Joseph Brant and to the Mohawks of the Six Nations – including a romantic liaison with a female relative of Brant's – earned him the contempt of the government of Upper Canada. Jones had worked extensively for the government during the 1790s, but after 1800 never worked for them again.

Typically, our earliest Ontario surveyors made deals, influenced treaties, and

Right: Sculpture commemorating David Thompson and his Métis wife, Charlotte Small, at Invermere, British Columbia.

advised on the creation of reserves. They established boundaries that were seldom in the interests of those being displaced or confined. Indeed, it was Ontario surveyors, acting aggressively on behalf of the federal government in the campaign to annex Manitoba, who triggered the Red River Rebellion in 1869, which resulted in nearly two decades of hostilities between the Canadian government and the Métis. Some might point out that it was the Métis themselves, under Riel, who sparked the conflict by executing one of those surveyors, a belligerent and racist Orangeman from southern Ontario, named Thomas Scott. (Having put Scott before a firing squad, Riel's provisionals placed his apparently-lifeless body in a coffin, realizing only when moans began emanating from the box, that they had nailed him in alive). In all, the conflict would lead to the slaughter of the Métis at Batoche in 1885, the hanging of Riel, and more than a century of embitterment on the part of those robbed of their land in the name of Confederation. The outcome at Batoche was of course but one element in the long campaign to subjugate the First Nations in the interests of settlement. It is eminently worth noting that an entire battalion of 55 Ontario surveyors went west with the Canadian army and militia in 1885 to finish off Riel once and for all. It is also worth noting that the name of just one Ontario surveyor, Michael Deane OLS #12, appeared on an 1885 petition to Sir John A. MacDonald, begging for clemency for Riel. A year later, Deane was present at the founding of the Association of Provincial Land Surveyors, in Toronto, and was present again in 1892, when the AOLS was incorporated.

Some of this was on my mind on a day in September 2015 as I talked by phone to the Espanola surveyor Bob Halliday about Dane Wandabense of Manitoulin Island, the first, and for many years the only, Aboriginal surveyor in the AOLS (Dane's career is outlined on page 156). I admit that, at the time, I found it ironic that a Native man should have become a surveyor at all during the 1930s and '40s, a period when surveying was frequently a government-sponsored activity and when government-promoted discrimination against the First Nations was still being openly practiced in Ontario. The residential schools were going strong at the time, and Native "Canadians" could neither vote in an election nor own land. They could, as Mr. Wandabense proved, go to war and fight for the values of an ostensibly democratic Canada.

Below: Surveyor Michael Deane, seen here during the late 1890s, was the only OLS to sign the petition begging clemency for Louis Riel after the Northwest Rebellion in 1885.

Shortly before talking to Bob, I had been reading about the work of Charles Rankin, who surveyed numerous roads and townships in Grey and Bruce counties and eastward into Muskoka. In 1854, Mr. Rankin was present at the signing of Treaty 72 that clearly divided present-day Saugeen Township on Lake Huron into "Indian" land and settler land.

However, when Rankin conducted the first official survey of the Sauble Beach shoreline and adjacent property two years later, he would *seem* to have violated the terms of the treaty by marking out an extra 2.4 kilometres of shoreline for the settlers – shoreline that not only belonged to the Saugeen First Na-

Most northern survey crews during the late 19th and early 20th centuries included First Nations guides, paddlers and axemen. This crew, working the Northwest Angle at Lake of the Woods, circa 1910, incorporated at least three men of Aboriginal descent – men who were reportedly better respected by the surveyors and crews than were the First Nations, generally, in the broad scheme of colonialism.

tion but had been vital to their fishing over the years. Today, the 160-year-old discrepency has become an issue between the Saugeen Band and township politicians. One Canada Lands Surveyor familiar with the dispute commented that Rankin "may simply have made a mistake, or the settlers misinterpreted his survey." Another veteran surveyor who has been involved in the settlement of numerous land claims, added that Rankin almost certainly did not carry out the survey himself but assigned the work to underlings, and that the discrepancy between the Township survey and the Reserve boundary was probably due to changes in the water line over the years and the acute angle at which the Reserve boundary met the shore of Lake Huron.

"Which is probably correct," says the surveyor Talson Rody, who has worked on several land claims over the years. "The problem is that the treaties and survey documents often didn't make fine points about water levels, and so on, clear to the First Nations – sometimes didn't make anything clear. And of course there were mistakes made in the interpretation of the treaties and surveys.... One thing I've noticed is that, for some reason, most of those mistakes were made to the advantage of the settlers and the detriment of the First Nations."

In May 2015, I had a long conversation with Julia Meldrum Smith, who in her role as a Canada Lands Surveyor has worked on the Akwesasne Reserve near Cornwall, and as a reviewer for the Association of Canada Lands Surveyors has done survey reviews on a number of other First Nations lands. She had recently returned from a national surveyors' conference at which reserve lands and the historic subjugation of the First Nations were extensively and keenly addressed.

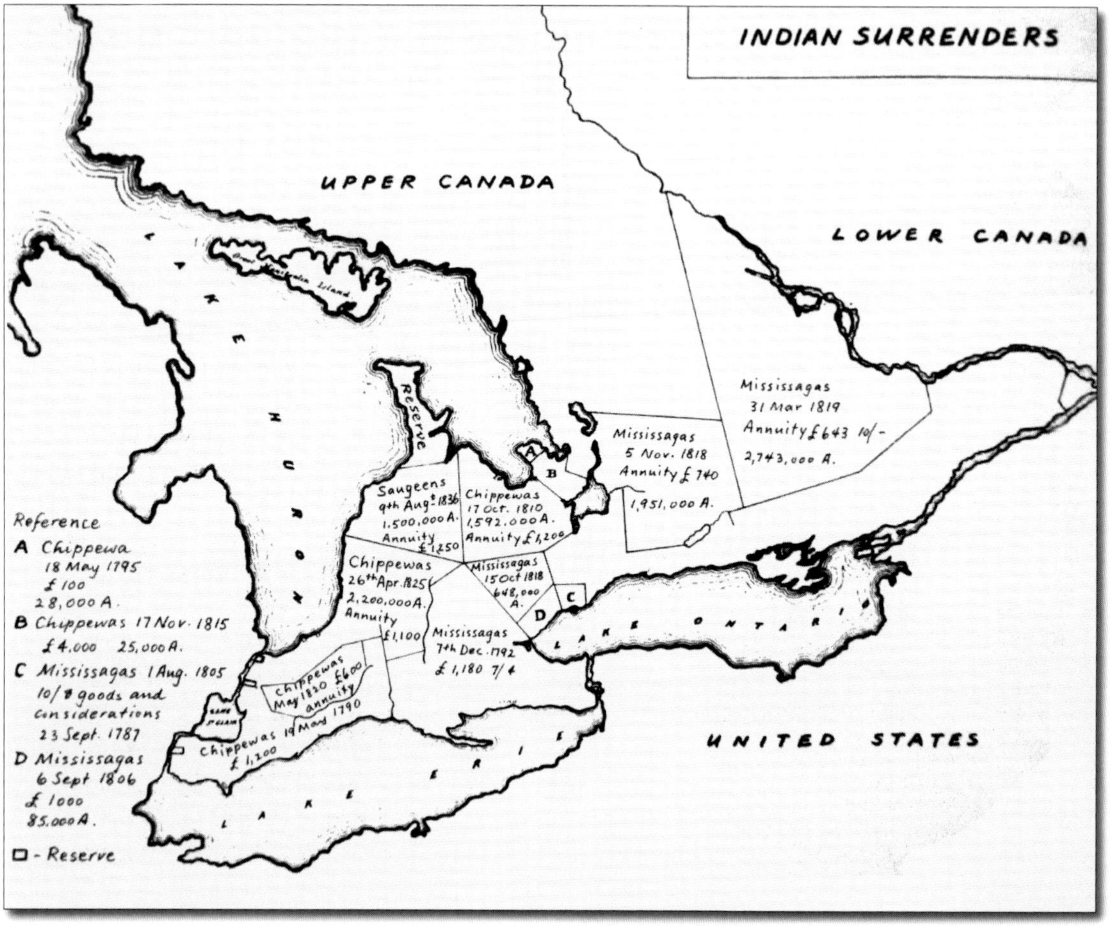

This mid-19th-century map shows tracts of First Nations land purchased by the Crown between 1787 and 1836 – all at prices ludicrously favourable to the government of Upper Canada and to European settlers.

"One of our speakers," she said, "was Gordon Bluesky, the lands and resources manager at Brokenhead Ojibway Nation, in Manitoba. His focus was the huge difference between how the First Nations understood the treaties, which were of course written in English, and how the Europeans understood them. Today, people are inclined to think, Oh, the First Nations were simply naïve in their negotiations. But it wasn't naiveté – the early tribes were in fact very savvy; they just happened to have an entirely different philosophy and understanding of land – in some ways a much more advanced understanding than the Europeans.

"In the Native philosophy," Julia continued, "you didn't own land individually – you couldn't! It was land! It was the planet! You lived on it; everybody in the tribe had the same freedom on it, the same sense of its value to the community, the same sense of responsibility for it. The tribes' attitude to the Europeans was: *Yes, sure, you can be here with us, we can cohabit the land according to the treaty.* But they were not negotiating to *exclude* themselves from the land in the way that they eventually got excluded. Gordon had a memorable analogy – he said, 'You think you're coming up with a roommate agreement, and the next thing you know you're locked in the shed on your own property.' As Canadians, we're not used to thinking of our history in that way. We like to think we didn't get our country by force. But we did. And surveyors were a part of the process."

Julia went on to say that because surveyors are frequently involved in land claims and occasionally in other surveys that concern First Nations, it is important that they understand what happened between the cultures, how the land was divvied up. "How can we possibly have a rounded and fair-minded view if we don't know the history?"

In all of this, she said, she felt changes for the better were taking place. "We're listening now," she said. "We're learning; we still have a ways to go."

Meanwhile, the surveying community itself continues to undergo big changes. Beyond increased sensitivity to land and environmental issues, the shifts, as we

Right: The famed founders of the MMM Group, (left to right) Oscar Marshall, Harold Macklin and Patrick Monaghan. Below: A group of Ontario surveyors visited the newly built Toronto SkyDome during the Association's AGM in February 1989.

have noted, are about advances in technology and about the perception of a shrinking community of surveyors. But they are also about the changing nature of competition among firms and the make-up of the firms themselves as more and more independent offices are acquired or absorbed by growing companies. Amalgamation, as Joe Young sees it, is partly about "economies of scale and an expanded breadth of services" and partly about geographic reach and access to markets. Mark Tulloch acknowledges that his firm settled in Huntsville because the town was and remains central to a number of major highway developments. During the past few years, Tulloch Inc. and J.D. Barnes Limited have each acquired five independent offices in Ontario, giving Tullochs a total of ten branches, J.D. Barnes a total of twelve.

"If things keep going the way they are," says Andy Shelp, "in ten to fifteen years there will be ten big survey firms in Ontario and a scattering of outliers – and that's all. That's my opinion. Barnes and Tulloch aren't going to be getting any smaller."

The same could be said for another of Ontario's mid-century firms, the MMM Group, perhaps our foremost example of how a once-compact survey firm has expanded far beyond its earliest interests and ambitions. Founded in 1952, when partners Oscar Marshall and Harold Macklin teamed up with Patrick Monaghan, the company's original aim was to serve the country's growing pipeline industry. The event that established the firm's ascendant reputation is said to have been its selection as chief consultant for the development of E.P. Taylor's celebrated model suburb, Don Mills, in east Toronto.

During the mid-1970s, MMM joined with two Montreal-based companies to create a branch they called Cansult, which pursued engineering and survey opportunities in Saudi Arabia, Oman and Abu Dhabi.

Within a decade, the firm had evolved into an industry juggernaut, chiefly in the engineering field, involved in everything from suburban development to international infrastructure projects, to the construction of skyscrapers, stadiums and airports. Among the company's better-known Canadian endeavors were the construction of the CN Tower during the early 1970s and the building of the SkyDome (now the Rogers Centre) during the late 1980s. Maureen Mountjoy worked for the company as a surveyor during that period. "One thing they had to make sure of with the tower," she says, "was that it didn't deviate even a centimetre in its vertical rise – anything more than that could have been a disaster." Likewise, at the SkyDome, the firm's surveyors had to be sure the 11,000-ton retractable roof was built, and stayed, within a hair's breadth of the specifics of its design.

By the beginning of the 21st century, MMM had seven offices in Ontario,

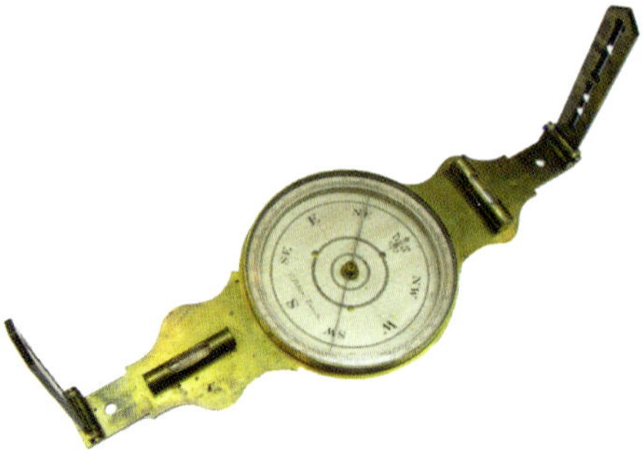

A brass circumferentor or sighting compass made in Toronto, circa 1880, by Charles Potter, a King Street manufacturer of surveying and engineering instruments.

eleven in other parts of the country, and dozens of affiliates throughout the world. Its industrial apotheosis occurred in 2015 when it was purchased by WSP Global, one of the world's largest engineering and consulting companies, with 32,000 employees in 500 offices in 40 countries – a long way from the firm's first small office on Merton Street in Toronto.

SUCH STORIES OF EXPANSION on a global scale are an equally long way from the toasty little archive in the AOLS offices where, as I have noted, I often worked late into the night on the book you hold in your hands. Sometimes, in the wee hours, under the glow of the fluorescent lights, lulled by the hum of the heating or ventilation systems, I'd get tired of browsing books and transcripts and the internet, thinking about boundary-making and technology, and for a break would get up and walk around the office, examining the art, the posters, the old photos, the maps, the certificates, the honour rolls and heraldry. Or I would descend the stairs into the basement, make a cup of tea in the lunchroom, have a snack, read a magazine. Occasionally, as a diversion, I would go back out into the hall and, inspired by curiosity (and perhaps the suggestion of illicit activity), would open a secret steel door with my pass key, which led to where the Association's collection of antique surveying equipment is stored. There, I would open one or more of the heavy wooden cases, examine the treasures, and sometimes lift an item from its felt-lined resting place. On a night in May 2015, on an impulse, I took an old wooden tripod from the room, set it up in the hallway, and bolted a transit to its head – and took a look to see what I could see down the hall; and for a half-hour thereafter fussed happily with old theodolites and compasses and circumferentors.

It does not take much under such circumstances to get the imagination churning. And, that night, back at my desk, in a remote part of my skull, I invented a surveyor, a time-traveller, who I envisioned ghosting through the pages of our book: along the trails and waterways of Upper Canada; a mapmaker, a treaty-maker, an agent of the Empire; student of the transit and chain; confederate of Holland, Wilmot, Unwin; in attendance at the founding of the AOLS; player in the great survey of 1900; reveler in the glories of the forest; seasoned by the hardships; builder of the province; of townships and farms; roads and railways; and, now, into the 20th century, of highways, power dams, cities; a survivor of wars, depression; now riding the boom years; the mining claims, the Seaway and freeway; the pipelines and airports and suburbs.

By 1970, our traveller, like all young surveyors, has acquired a degree; then a desktop, a laptop, a total station. By the turn of the millennium, he is no longer a "he" but a "she" – an Erindale or Ryerson or York grad; perhaps from a far country; rookie at articles; master of technology; wiser than we might have imagined; more prescient about the future; more knowledgeable about the past; perhaps with a sense of righting old wrongs, redrawing old maps.

As the reverie deepens and then resolves, we see our young traveller moving across a landscape, a province – forest, fields, flood plains; arriving, finally, at a height of land, a precipice; behind her the familiar, before her the unknown. She walks to the edge, confident in her knowledge and in herself. She is carrying neither a transit nor total station nor GPS…nor electronic device, nor screen of any kind. She is carrying, rather...

a book
an old-fashioned thing,
a throwback to the time in which
her travels began.
She opens it, absorbing at a glance
its photos and illustrations and maps.
She examines, ponders, laughs…

It is not a textbook,
not an instruction manual,
nor an analysis;
not a history book either
but a storybook, whose lives are her lives,
whose stories are her stories
... these stories;

whose celebration is her celebration.
She stands at the edge of the world.
We stand beside her
and with the turning of a page
begin to read:

Early AOLS members on the steps of the east entrance to the newly constructed Ontario Provincial Legislature, Toronto, circa 1900. Front row, from left: *James Dickson, Elihu Stewart, Alexander Niven, George Kirkpatrick, Willis Chipman.* Back row, from left: *George Ross, Peter Gibson, Otto Klotz, J.W. Tyrell, C.A. Jones, Thomas Fawcett, A.J. van Nostrand.*

CHAPTER ONE

JOIN OR JUMP... GOD SAVE THE QUEEN

IT MAY NOT HAVE BEEN THE FINEST HOUR in the history of Ontario surveying, but it was arguably the "funnest." And on the afternoon of February 24, 1889, not a single one of the 48 surveyors who had gathered on Front Street, with a dozen of their wives, was going to let the frigid Toronto temperatures or the piles of horse poop in front of city hall get in the way of his or her enjoyment of the day. The unlikely event for which the surveyors had assembled was neither a memorial nor dedication nor the laying of a cornerstone – all of which were routine for surveyors – but a sleigh ride, of all things, an event pitched by its organizers as an "opportunity to view some of the contemporary marvels of surveying and engineering in the provincial capital." More realistically, it was a bit of a caper, a couple of hours of fun and recreation. The event had been laid on by the city fathers, who had promised the surveyors everything from the newest model sleighs, to seat warmers and blankets, to cocoa and biscuits at the popular St. Clair Bakery. And as the day progressed, things were only going to get better. Three hours hence the sleigh riders would be unbuttoning and unburdening at the gracious old Reform Club on Wellington Street. Dinner there was to include Belgian bisque and French-seasoned capons and English trifle, with the usual sides of laughter and music and the best wine and whisky the city could offer.

As they stood on Front Street awaiting the sleighs, every member of the party was bundled to the ear lobes in furs and muffs and scarves. Some were carrying silver flasks of scotch in their inner pockets, and more than a few smoked bankers' cigars, bought for five cents a light at Benny's Tobacco Shop on Church Street.

The men had spent the morning a few blocks to the north on Richmond Street

Peter Gibson of Ontario's "first family" of surveying was a steadying voice at the founding of the AOLS. Below: Thomas Fawcett fought Riel and Dumont at Batoche during the Northwest Rebellion, in 1885, and returned, injured, to help found the APLS in 1886 and the AOLS six years later.

in the upper chamber of the drafty old Royal Canadian Institute – listening to papers, arguing policy, shivering by the fireplaces in a building that, like so many in Victorian Canada, lacked central heating. It had been noted that morning, in jest, that it was only because they were surveyors, accustomed to the elements, that they possessed the fortitude to carry on, hour after hour, in any state of discomfort, at any temperature, under any level of illumination, including the purgatorial light of the dank Canadian Institute.

Their tone may have been somber, their hatbands sooty, but the surveyors themselves were nonetheless thrilled about the success of this the fourth annual meeting since the founding of the Association of Provincial Land Surveyors in 1886. They all agreed: this was the meeting that counted; this was the watershed. Their secretary, the wily and impulsive Willis Chipman, had assured them (based, he claimed, on research) that if this or any other organization could survive four years, it could survive a hundred, perhaps a hundred and fifty, although nobody could come close to imagining what an association of surveyors might look like in the gaudy fogs of the year 2039. There were those who said that by that distant date horses would be gone from the streets, trains from the tracks – that people would be travelling in self-propelled carriages, perhaps even flying machines, all impossible to imagine!

What's more, the membership had risen to a heady 122, up from just 25 during the association's inaugural year.

Best of all, there had been low-key talk of incorporation, which would be the final step in the creation of professional independence and self-regulation for the surveyors of the province. Undeniably, there was resistance. Zany old Spry, for example, curmudgeon to the end, was still on about the death of surveying, the impoverishment of the practitioner, the lack of available work. Three years ago he had assured the gathered surveyors that if he "had a horse" he could do all the surveying that needed doing in two counties near his home in St. Mary's. The Toronto surveyor John McAree is said to have muttered at the time that "if William Spry had a horse, someone would undoubtedly have tied him across it by now and slapped its backside."

What really irked him, Spry had protested, was the outlandishness of being expected to fork over three dollars to join "apples," as he referred to the APLS, and then two dollars a year, about half a day's pay, to belong to an association that was likely to be as droopy as a eunuch when the hard demands of reality set in. Peter Gibson, the most respected urban surveyor in the province, and a

The Canadian Institute, at Richmond and Berti streets, in Toronto, hosted a transformative meeting of the APLS (soon to be the AOLS) in 1889. Below: Toronto's first city hall at Front and Jarvis streets, is now part of the St. Lawrence Market, where it is visible within the market building.

councilor of the association, was himself somewhat reluctant. While he believed heartily in the organization, he could not see the need for the creation of a corporate body that would control everything about surveying from standards and entrance requirements to articling, education and discipline... and, what's more, would get to rewrite the Surveyors Act on its own terms, to tell the government what to think about surveying. Gibson demanded to know what would happen to surveyors who chose to remain independent of the official apparatus. Would they simply be cut off from their historic standing, denied the right to make a living? The question provoked a lull that was broken only when the veteran Elihu Stewart rose, reflected for several seconds, and admitted that that was exactly what would happen to those who chose not to support the association. "Join or jump," he said. Those were the choices.

MERCIFULLY, THEY WERE CHOICES that did not have to be made on the spot. Indeed, at the forgiving hour of 2 p.m., the surveyors abandoned their wrangling, took the horse-drawn streetcar down Church to Front and then east to City Hall, where the sleigh ride would begin at three. The men had barely arrived and located their wives when the sound of harness bells and the muffled thud of hooves on snowpack came rolling down Jarvis Street, and, from behind the building, the satin-black horses appeared at a trot, pulling two-seat cutters as bright and sleek as chariots. If the presence of twenty Arabian horses and ten shiny sleighs was a thrill for the surveyors, it was no less a spectacle for the shopgirls and newsboys who were afoot in the street in the ripening afternoon.

Within seconds, the party was aboard, blankets across their laps, collars up, hats pulled snug over ears. At a signal from the hindmost driver, the teamster gave a whistle, *easy out*, and off they went with a cheer – a boisterous parade, west on Front to University and north up that wide and turd-strewn thoroughfare toward the accumulating pile of pink sandstone that by 1893 would be the new Ontario Legislature. A handful of the surveyors, including Peter Gibson, knew the place well, had surveyed the property, while others had applied their transits and tapes to making sure the building itself was sound and true. The joke in the Toronto *Globe* had been, "If only it were possible with a transit and line to assure the soundness and trim of the government inside."

From Queen's Park, the cavalcade pranced north up Avenue Road to St. Clair, east to the Rose Hill Reservoir, and

George Kirkpatrick, seen at left, was Director of Surveys for Ontario from 1891 to 1917, as well as the first president of the APLS and a force in the founding of the AOLS. Right: The third Ontario Legislature, on the north side of Front Street in Toronto (where the CBC building now stands), hosted the earliest meetings of the APLS during the late 1880s.

angled southeast from there to the Gerrard Street Bridge. There, a fine new replacement span, built of arched iron and concrete, was in the early stages of bridging the Don River.

Finally, at dusk, the sleighs skidded to a halt outside the Reform Club, every rider happy to be heading indoors.

What ensued was a night that, in the words of the association's first president, George Kirkpatrick, "would mean more to the surveyors' awareness of province-wide issues, more to their sense of mutual support, than would two weeks of meetings, motions and debate."

When dinner was done, the entertainment began – which is to say the self-entertainment, hardly an item of which was allowed to pass without an accompanying toast to her Majesty, the sturdy little monarch, who inspired now a rousing rendition of "God Save the Queen."

Between bouts of genuflecting to Great Britain, to colonial subservience (including a mid-evening chorus of "Rule Britannia"), various surveyors played piano or sang Victorian tear-jerkers the likes of "Never Take the Horseshoe from the Door" and "Just Before the Battle, Mother." Against the momentum of the evening, and surely against the preferences of the surveyors, one young surveyor took the floor midway in the festivities and delivered a fifteen-minute lecture on "amateur photography," a talk reported to have been "greatly condensed," given the broader spirit of the occasion (and perhaps the grumbling of those in attendance). But they were quickly back to singing, eventually closing the place down with a rendition of "Auld Lang Syne," accompanied by a double toast of cognac, and, in case anyone had missed it, a second singing of "God Save the Queen."

It was all in stark contrast to the association's first get-together, three years earlier, in 1886, at the old Ontario Legislature on Front Street, a building that for draftiness made the Canadian Institute feel "cozier than a stool by the stove," as one observer put it. The Legislature had been allowed to deteriorate to the point where the 36 surveyors who had gathered were obliged to wear their overcoats until the building warmed up at midday. While hopes were high among those who showed up, there had been no official welcome from the city, no entertainment or *joie de vivre*, no wives present, and most certainly no secure sense that the organization would get properly launched. A couple of previous attempts to form a surveyors' association had been gloomy failures. Many of the surveyors themselves were not in terribly good trim in 1886. Two or three, including the respected Thomas Fawcett, had recently returned injured from their efforts on behalf of the Dominion in crushing Riel at the battle of Batoche (another had died of bullet wounds in the same fierce encounter). Several had reached the meeting, having had the stuffing kicked out of them by the blizzards that had gripped Ontario with uncharacteristic ferocity since December. Trains had frozen up. Horse carriages had been buried in drifts along the roads. One surveyor was said to have set out snowshoeing to the train, thirteen miles away, but had been forced by the cold to turn back. It was an age of diphtheria and scarlet fever and spinal meningitis, and no one was immune

when the temperatures dropped. Many more surveyors, including a dozen or so who had been instrumental in opening Ontario to settlement, were simply passing from the scene, either dead or at "death's creaky door," as Willis Chipman characterized their position.

As for the 36 faithful who had arrived, what inspired them, beyond their toughness and pride, was the very real need for an association that would bind them into a political and scientific force. Provincewide, there were said to be 200 of them, but they were scattered to a degree that deprived them of any collective strength, or even an awareness of the issues that each faced in his corner of the province.

"Union and cooperation are magical watchwords in other branches of work," declared George Kirkpatrick of the Department of Crown Lands at that first meeting. "The time has come when, if we wish to succeed in elevating the profession of land surveyor, we must unite, and no longer remain isolated and unknown to one another." He reminded those gathered that "a general advance in knowledge" was taking place, and concluded with the warning: "If we do not as a body advance with the rest, we shall find ourselves hopelessly overmatched and dragged down from the position which a well-educated and practical set of men should occupy in the community."

On a more detailed level, Otto Klotz observed that formal association would give surveyors the clout to push for revision of outdated legislation pertaining to the profession. Tom Fawcett declared that association would "elevate standards," both for the entrance to and practice of surveying. The famed wilderness surveyor, Alexander Niven, believed union would bring discipline and new possibilities for "professional development."

Others asked simply: Why should we not have an association, when the doctors, lawyers, engineers and bankers all have one? "Each of us knows in his heart," said Niven, "that his work is just as important to the province's well-being as that of anybody in any other profession." John McAree had the jam to point out that the "damned province wouldn't even exist, couldn't have been established," had it not been for an army of hardy transit-bearers who in many cases risked their lives laying out districts, townships, municipalities, farms, roads; who fought snowstorms and sometimes hunger (not to mention blackflies, black bears and black moods) to make the province "ownable"… "occupyable"… "exploitable." That was the European way: divide up the land so that it could be managed and consumed. It was in that spirit that surveyors

Willis Chipman • An Orchid against Disease

There are no trophies awarded, or parades held, in honour of those who design and build our sewers – or slog through them in the nasty business of "human waste management." Yet these anonymous public servants make an immeasurable contribution to the prevention of disease and the livability of our towns and cities.

If there were a holiday devoted to such folk it would, in Canada, almost certainly be named for the pioneering surveyor and engineer, Willis Chipman.

If Willis is remembered at all these days among the majority of Ontario surveyors, it is for his drive and inspiration in the founding of the Association of Ontario Land Surveyors in 1892 – and perhaps for his inauguration of the organization's annual Veterans' Dinner, a century-old event that to this day is a signature celebration at the association's AGM.

Born in 1855 in Athens, Ontario, northeast of Kingston, Chipman was just 29 when, as chief engineer in nearby Brockville, he recognized a crucial need for the separation of storm and sanitary sewers in that town – and for a proper sewage treatment system. At the time, outbreaks of cholera, scarlet fever, typhoid and smallpox were common in Canada, where effective sewage treatment was in its infancy. "However beautiful and healthful our town and however pure the breezes of the St. Lawrence," wrote Chipman in 1884, "we cannot ignore that we are accumulating in our midst all the elements necessary for the spread of preventable disease."

So successful was Chipman's sewer design in Brockville – so instrumental in the quelling of infectious illness – that by 1890 other towns and cities were clamoring for his services. In all, during the next fifteen years, he designed sewers for nearly every municipality of any size in Ontario (some 50 in all), as well as for 34 towns and cities in other parts of Canada. He studied and consulted on sewage systems in American cities as far afield as San Francisco Denver and Seattle.

Unlike many of the more rustic surveyors of his day, Willis was slim, dapper and clean-shaven, his hairless face a rarity in an era prior to King Gillette's invention of the safety razor in 1901. One of the few surviving photos of Chipman shows him in high starched collar and silk tie, and wearing a splendid white orchid on the lapel of his black suit (it has been suggested, in jest, that the flower was to dispel the odour of sewers, with which he was intimately familiar).

His nattiness carried over to his notebooks, the earliest of which, written during his surveys of several Ontario townships, are decorated with exotic, sometimes extravagant, hand lettering and insignia the likes of which would, these days, require a computer to execute.

Elected president of the AOLS in 1896, Chipman fought strenuously for improved education and apprenticeships for surveyors, as well as for more rigorous examinations not just in applied science and math but in Canadian history, geography and flora, and in English grammar. At the same time, he argued for improved maps of Ontario, pointing out that the rivers on many existing maps were often shown flowing into the wrong bodies of water – or that known bodies of water were omitted entirely.

In 1895, Chipman urged the Washington-based National Geographic Society to include Ontario surveyors in the society's forthcoming exploration of the Arctic. Rebuffed, he proposed an all-Canadian exploration, a mission

Left: *During the 1890s and early 1900s, Chipman designed closed sewers for cities across Canada and into the U.S., including the one illustrated here in Oakville, Ontario.*

Laying the Sewers at Oakville, *oil on canvas, Moma Markovich, 1966. Goverment of Ontario Art Collection,* Archives of Ontario *101216*

eventually accomplished between 1904 and 1908 by Captain Joseph Bernier, who piloted the Canadian Coast Guard vessel *Arctic* deep into polar waters, claiming much of the territory for Canada. From the boat's decks, members of the Canadian Geological Survey took bearings, sketched maps, made notes, disembarking occasionally to place survey monuments in exposed soil or rock.

In 1896, Chipman proposed a systematic exploration and mapping of northern Ontario, with a view to resource development – an initiative that was eventually manifested in The Great Survey of 1900.

Chipman was an accomplished ornithologist and historian, and an indefatigable writer, who in mid-career penned the first of some 250 short biographies of Ontario surveyors – many of which can be found on the AOLS website.

Chipman died in 1929 in the presence of several of his children from three marriages, and of his third wife, Edith. It is said that he never fully recovered from the loss in 1883 of his first wife, Angeline, the mother of his two oldest children. He had gone west in early spring that year for the money he could earn surveying for the railway companies, and had returned to find Angeline recently dead from one of the very illnesses that he would spend much of the heart of his career working to eliminate through improved sanitation. It was perhaps in her death, and in his determination that such tragedy should not befall others, that Chipman forged his indelible legacy in the service of human health.

He is buried in Old St. Peter's Cemetery in Brockville.

Below: *Retired Brockville surveyor, Ken Wiseman, one of a succession of heirs to Chipman's original surveying practice and field notes, stands by the visionary surveyor's grave in Old St. Peter's Cemetery in Brockville.*

such as Rankin and Jones and Wilmot – and McNish and Burwell and Salter – had cut up the territory, so that governments, whether local, provincial or federal, could piece it off to those who were in the process of building the country.

Thus it was that, during the late afternoon of February 23, 1886, with the light dying at the windows, a motion was made, a vote was held, and as quickly as that the Association of Provincial Land Surveyors was brought into being.

In summing up and encouraging as broad a membership as possible, Willis Chipman delivered to the surveyors the following rather eccentric *au revoir*: "To the very clever surveyor we would say that you will undoubtedly become a 'crank' unless you occasionally rub up against your professional brethren and get the 'corners' polished off. To the 'rusty' brother, we think there is no better way of brightening you up than by joining what we are confident is the permanent organization of the Association of Provincial Land Surveyors of Ontario."

FAST FORWARD to the radiant evening at the Reform Club, three years hence – an event that left members brimming with a sense of unity and significance and optimism. Some were energized to the point where the following morning (not too early), they introduced to their proceedings a stirring new level of debate – specifically a discourse on whether Ontario should abandon Roman and imperial measure and, instead, apply metric measurement to surveying and all other scientific endeavours. Nearly a century would pass before metric became the standard in Canada. However, the debate initiated a series of discussions and papers that, at annual meetings during the next few years, brought a lively new standard of cerebral and professional conversation to the association's get-togethers.

One such discussion a year later introduced the idea of a wholesale, integrated survey of far northern Ontario – a survey more thorough and organized than the haphazard surveying that had been imposed upon the southern part of the province. Another exposed the need for a topographical survey of Canada so that the pathetic state of mapping in the country could be rectified.

James Whitson's paper on toughing out a timber survey, deep in the bush, in winter, elicited responses from the surveyors that plumbed the depths of their sense of adventure and survival. A surveyor, Whitson told them, had no place in the bush "if he could not snowshoe, pull a toboggan, and drive a dog-team." As for the dogs, a trio of them should be expected to haul up to 450 pounds as far as 35 miles a day, on good trail. "A dog's daily ration," said Whitson, "should consist of a pound of cornmeal mixed with a third of a pound of lard – or, otherwise, three pounds of fish, on which they would not work as hard as on the corn and pork fat."

The lecture touched on fires and blankets and wall tents – and stoves of "light Russia steel," which Whitson said any caring government should supply. James Dickson responded that he and his men had been perfectly happy in the north with no stove at all – a contention that brought James Sewell of Port Arthur to his feet, arguing that the best way to stay warm was to build a good-sized bonfire between a pair of sleeping tents whose open doors faced the flames. George Kirkpatrick claimed to know a surveyor (unlike the pantywaists before him) who lit no fires at all in the depths of winter, but rather crawled into his sleeping bag in the snow, closed the drawstring over his head and was as warm as a bunny through the night.

A question arose over what sort of lamps shone brightest for night observation. According to the outgoing presi-

Few items were more important to early wilderness surveyors than were the canvas wall tents that protected them from the elements. The advertisement at right appeared in the printed proceedings from the Association's inaugural AGM in 1892. As noted in the ad, John Milne and Co., of Toronto, sold surveyors everything from tents, stoves, cookware and beds to axes, lanterns and brush hooks.

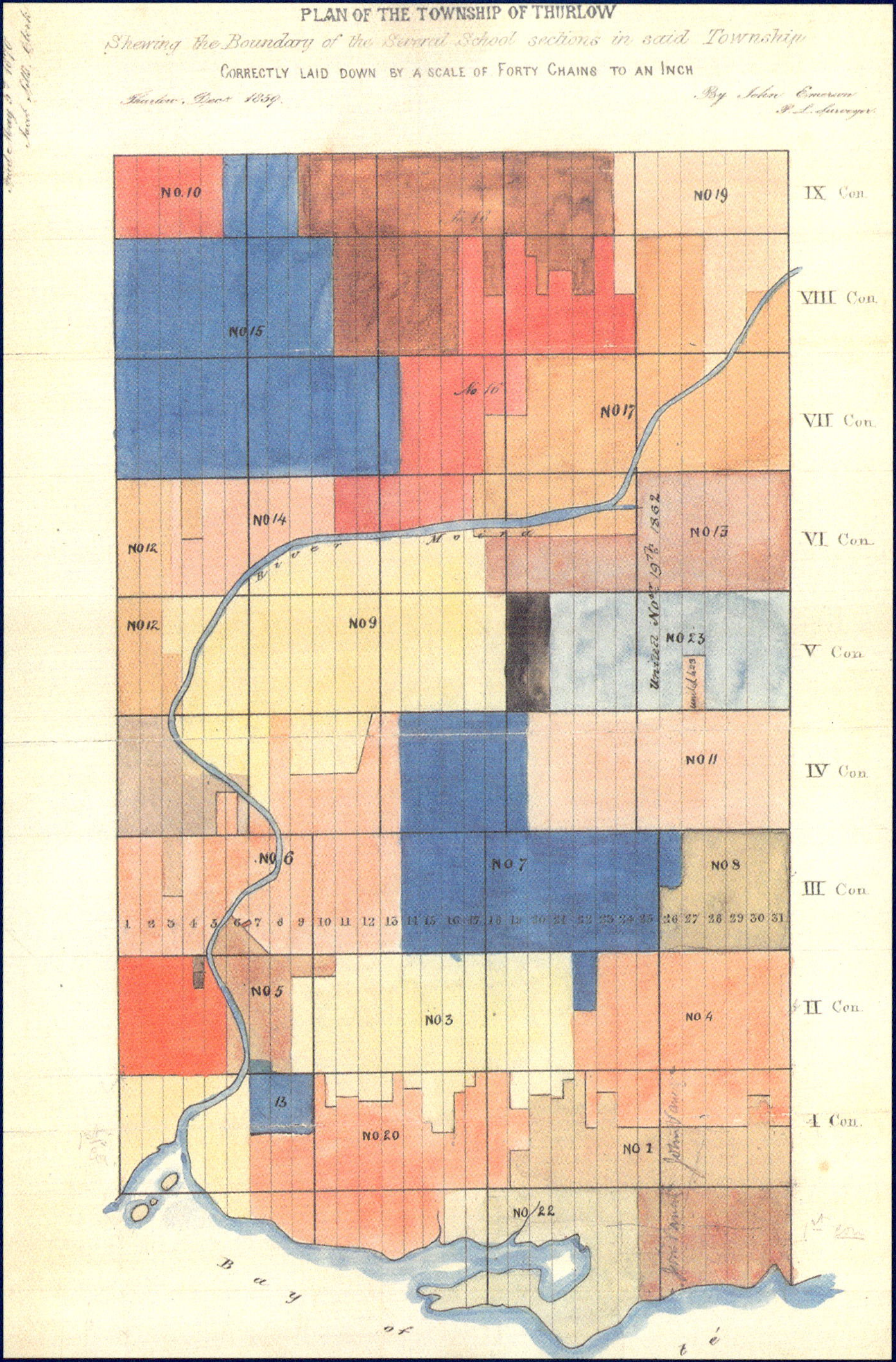

Thurlow Township spreads north and east from Belleville, the township's biggest municipality, located where the Moira River, shown on the accompanying plan, flows into the Bay of Quinte, on Lake Ontario. The 1859 plan, drawn by Provincial Land Surveyor, John Emerson, depicts in colour the areas serviced by each of the township's 21 schools (20 public, one Roman Catholic). In all, the schools were attended by nearly 1,200 children from across the township's 15 square miles.

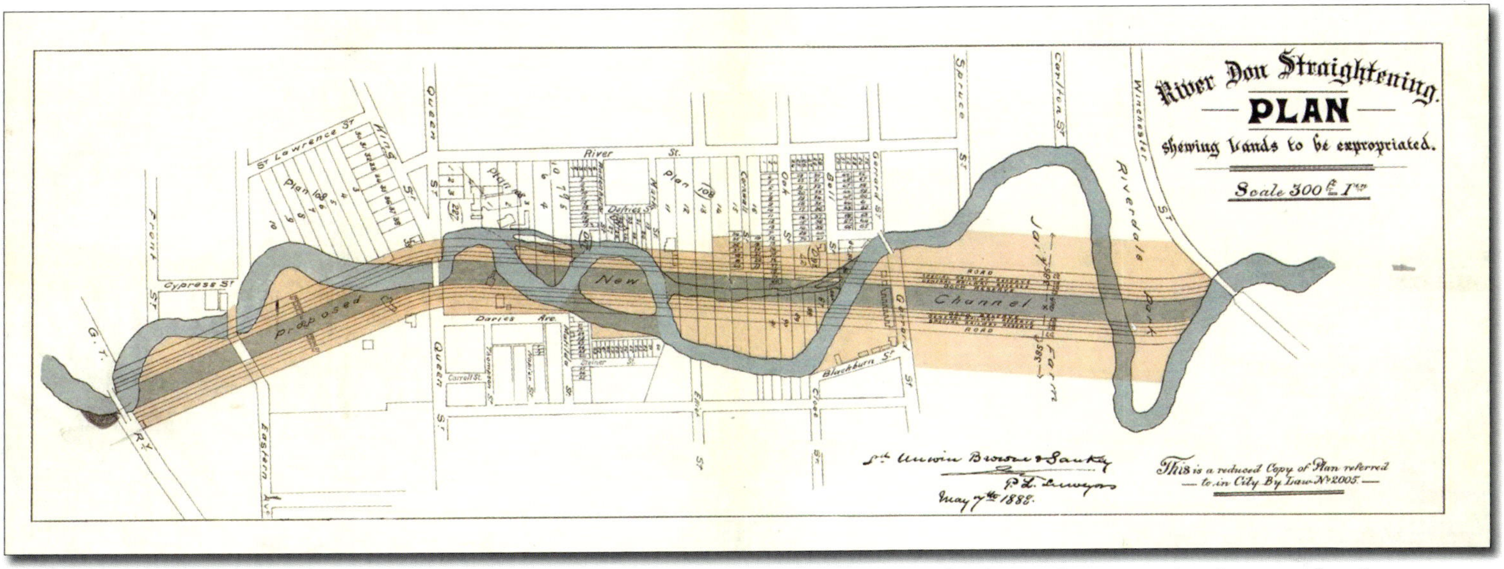

In 1888, the Toronto firm of Unwin Brown & Sankey drew up the accompanying plan to straighten the once-serpentine Don River between Winchester Street and what is now Lakeshore Blvd. in Toronto.

dent, Villiers Sankey, it was a brick soaked in coal oil. Asked how to keep a theodolite functional at 50 degrees below zero, George B. Abrey recommended rubbing a little badger oil on the spindle,

Charles Unwin was among the founding members of the AOLS in 1892.

because it stayed soft at any temperature. Under no circumstances should a petroleum product be used.

As compelling as such discussions were, the meeting of 1891 was dominated not by intellectual debate, or by discussions about the practice of actual surveying, but by one concern and one only: the drive toward incorporation. In the minds of the masters of the profession, the time had come – dissenters must be brought onside.

In the end it was easier than anyone might have thought – in large part because of well-reasoned arguments from a handful of the most respected surveyors in Ontario.

Peter Gibson again voiced his concern over what would happen to a surveyor who refused to join. And the answer, as ever, was that he would no longer be allowed to practice in the province.

What was needed immediately, it was agreed, was the drafting of an act of parliament that could be presented to the government for passage.

Late in the afternoon of the meeting's second day, the Belleville-area surveyor William Aylsworth rose and presented a motion: "that it is desirable to have the land surveyors of the Province of Ontario incorporated… and that this Association should take steps to have an Act therefore prepared."

With the motion passed, it was agreed that the documents and constitution should be drawn up. Villiers Sankey would head a committee in that direction. Meanwhile, the surveyors of Ontario, particularly those not in attendance, should be put on notice that surveying in the province was about to move forward. And that they should prepare to move with it.

In 1889, James Dickson, OLS #102 (top left), was appointed to an Ontario Royal Commission the aim of which was to establish Algonquin Park, which opened in 1893. To this day, the park allows regulated logging activity, meaning that the massive white pines depicted in the 1906 photo at left are long gone. Since the 1930s, when decent roads reached the park, it has also been a favourite recreation place for visitors seeking a brush with the wilderness.

Major Villiers Sankey, who was central to the founding of the AOLS, completed a 17-year term as chief surveyor for the City of Toronto in January 1905, and died seven months later in a canoing accident on a surveying mission on Rainy Lake, in northwestern Ontario.

WHEN THE LOT OF THEM gathered again in February, 1892, it was to seal the deal: to incorporate and move ahead. By that time, George Kirkpatrick had taken charge of the act that had been drafted. It was he who would present it to the legislature for approval.

The document said among other things:

That the Association of Ontario Land Surveyors would take charge of its own affairs – which is to say, discipline and education and standards. That all Ontario surveyors would belong. That those who didn't belong couldn't practice. That those who didn't belong and practiced anyway would be fined 20 dollars the first time they were caught and 50 the second time (about ten days' pay). That the Association would have an executive consisting of a president, a vice-president and six councillors. That the executive would work to amend the Surveyors Act.

By the new terms, the Association could own up to $5,000 worth of real estate (this at a time when the price of a decent house in Toronto was about $1,500).

Once a year the Association would be obliged to hold a meeting, beginning on the fourth Tuesday of February. The meeting would include a dinner (the understanding being that the dinner would include lots of good wine and scotch, an evening of entertainment, and an exaggerated degree of genuflection to the Royal House of Hanover).

WHILE MUCH HAS CHANGED for the association over the years, much has not changed. Questions of standards, discipline and education still arise regularly. As do questions over fees, the future, and respect for the profession.

No question has ever been raised about the importance of the dinner. Or the good wine and scotch.

The energetic singing of "God Save the Queen" and "Rule Britannia" has tailed off.

Another dictate of incorporation was that each surveyor would be given a number. At the time, this was adjudged to be of little consequence. But as the years passed, the number a surveyor was assigned became less a mere jot in the chronology and more a kind of identity within a family.

Surveyors are a family.

Which is not as bad as it sounds.

Villiers Sankey said, "We are connected in our interests and standards and in our sense of mutual support. And in that we are brothers." In those days, there were no sisters.

Mr. Sankey spoke those words two years before incorporation.

At the first AGM after that watershed event, Sankey rose at the grand dinner and congratulated the surveyors on what they had become. And wished them well. Again, he referred to them as a family. "We have worked hard," he said. "We have been responsible to ourselves, and thereby to the province. We are at last in charge of our own affairs."

And he invited them to raise a glass.

Since 1991, former AOLS president, Jack Young, has planned and hosted the Veterans' dinner, an annual feature of the AGM, dating from 1916, when it was initiated to celebrate the veterans of the profession. Jack has also been an instrumental presence at the Senate Dinner, held annually at the AGM. One of the features of the event, as suggested by the glassware and sustenance in Jack's hands, is a series of hearty toasts to surveyors past and recently departed… and to whatever other milestones might come to mind for such remembrance.

Webb's Restaurant, site of the association's grand dinner during the AGM of 1894, was reputed to be the finest in Toronto. The poster seen above is the work of Frederick Lucas Foster, OLS #69.

1894

1967

1963

193

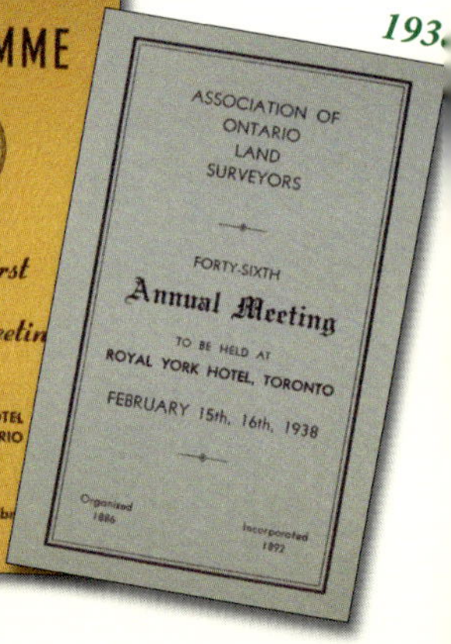

1909

1931

The AGMs

As much as any other aspect of the life of the Association of Ontario Land Surveyors during its 125-year existence, its Annual General Meetings – their spirit, the mood of those present, and the artifacts by which they are remembered – have been a telling reflection of the state of the profession and of the broader history of the times. For example, the accompanying photo of those who attended the AGM in 1909, during the boom years of the early 20th century, is pretty much a mural of prosperity: men in jaunty hats, with jaunty expressions, shoulders back and proud. The equivalent photo, taken 22 years later, is, alternatively, a study in the prevailing despair of the Depression: men with sour expressions, hats in hand, shoulders slumped. Likewise, the photo of the attendees of 1969, at dinner in the King Edward Hotel, all but zings with the excitement of the age.

1969

The roots of the tradition are unclear, but at each of the Association's AGMs, the ceremonial sergeant-at-arms assumes the identity of a historical or fictional character of his or her choosing. From top left: Ron Berg (1996) as the 13th-century Scottish warrior, William Wallace or "Braveheart;" Murray Purcell (1991) as a rather flamboyant Samuel de Champlain; Drew Annable (2016) as the 19th-century Ontario land potentate, Thomas Talbot; and Nancy Grozelle (2003) as the 15th-century French soldier, martyr and saint, Joan of Arc. The sergeant-at-arms opens each AGM with a ceremonial presentation of the Association's Standard Measure (seen above), a yard-long brass rule made in England in 1851, against which a surveyor in Upper Canada could check his rule or chain.

The AGMs

The surveyors' hockey game has become an anticipated feature of the association's annual meetings. The 2015 contest was played at Deerhurst Resort, Huntsville, where the goaltender for the white-sweatered team was the estimable Julia Meldrum Smith (seen above, foreground).

Below: The association's presidential chain of office was redesigned in 1985 to include the modern AOLS logo and the name of each president from 1985 on. The historic chain of office, created during the 1890s and bearing a plumb bob, a compass and an optical loop, appears on page 79.

Right: Veteran OLS Neil Simpson and his wife, Rose Marie, cut the rug at the Association's 100th AGM in Toronto in 1992.

Beaching the canoes and unloading them: The Great Survey of 1900 – Party #1, Lake Abitibi.
Archives of Ontario 10052450.

CHAPTER TWO

The Great Survey
The Great Clay Belt
The Great Agricultural Hoax

When the renowned pioneer land surveyor, Alexander Niven, of Haliburton, reported to the Ontario Department of Crown Lands in 1886 that he had discovered "land admirably suited to agriculture" northwest of Lake Timiskaming – had discovered hundreds of square miles of it – the officials of the department, including Surveyor General, George Kirkpatrick, could not have been happier. It was precisely what they and the rest of the provincial government were itching to hear. They knew there were metals and minerals in the north, knew there were trees by the billions, knew there was wealth. What they did not know and wanted to know more than anything was that there was *farmland* somewhere up there amidst the sprawling forests and rivers north of the CPR line. In their mind's eye (give or take a blind spot or two) they could see wheat, could see hay, could see vegetables; could see cattle and pigs and sheep. Could see *settlement* – could see it as surely as it existed on the fertile but increasingly crowded farmland of southern Ontario.

What they apparently could not see from their comfortably furnished offices in Toronto was the north's swamp and muskeg and moss; its bad drainage, its boulders, its granite outcrops, some as wide as the farms of Oxford and Essex counties. They could not see a growing season better suited to touch-me-nots than turnips.

Government ministers who should have known better envisioned the Northern Farmer of the Future, a stalwart of progress, driving his draft horses and hay mower in the lush heat of July, the sun haloing his straw hat. Unfortunately, they could not envision him in December or February, shivering by his inadequate

This enticing poster was produced by the Ontario Ministry of Lands, Forests and Mines, circa 1918. But its depiction of the prosperous northern farmer, in his tailored pants, matching shirt, and flawlessly polished boots, was an illusionary and misleading southern Ontario fantasy, nothing more. (The poster and its companion, shown on page 97, were discovered at the Aberfoyle Antique Market in the mid-1980s, by Lorraine Setterington/Petzold, executive director of the AOLS.)

Growth was sudden and unpredictable in mining towns such as Cobalt, seen here during the early years of the 20th century. Surveyors did well in such places, laying out townsites and developing silver and gold mines. In Toronto, government officials fantasized that agriculture in the north could somehow keep pace.

hearth – or in late May, knee-deep in muskeg, sinking slowly into penury and embitterment.

It perhaps goes without saying that where George Kirkpatrick and the Department of Crown Lands saw settlement, they also saw surveyors – in this case ranging over rock, across rivers, deep into evergreen forests, laying down base lines and meridians, plotting townships and concessions, farm lots and roads; saw them opening the territory for mines, lumber mills, rail lines, perhaps for a couple of those new-fangled hydro-electric dams that Thomas Edison had imagined into being (one had been started up recently on the Fox River in Wisconsin). Surely the rushing rivers of northern Ontario would be perfect for such dams.

Forty years later, Niven's pronouncement on the north's suitability for agriculture would be viewed in eye-rolling, if not laughable, hindsight. As would a subsequent decree by surveyors that the Little Clay Belt in the Timiskaming area was dwarfed by a much larger tract of agricultural possibility, the so-called Great Clay Belt, which extended 2,000 kilometres to the east and west, north of Lake Abitibi, into the unknown expanses of Ontario and Quebec.

Meanwhile, Niven's declaration became the unwitting inspiration for one of the more outrageous public hoaxes in Ontario history: namely, the well-publicized attempt by the Department of Crown Lands and the Office of Colonization to persuade would-be farmers that there were crops to be raised out there in snowy Xanadu, crops that could be had for the seeding and harvesting by pretty much any opportunist who decided to venture north.

In 1886, in aid of such seeding and harvesting, Niven and other surveyors were dispatched at the government's expense to begin charting the territory. They started in the near-north, the area around present-day Sudbury, Cobalt and Kirkland Lake.

During his first eight months in the wilderness, Niven laid out seven townships on what he continued to insist was "arable land the equal of any left unfarmed in the province." The following year he outlined eight more townships near present-day Haileybury – laid down their margins, leaving their division into lots to those who followed.

By 1890, a total of 24 townships had been charted on the Little Clay Belt, northwest of Lake Timiskaming; eleven more in Nipissing District north of Mattawa; sixteen in the District of Sudbury; and fourteen in Algoma, east of Sault Ste. Marie. Further west, seven had been plotted in Thunder Bay District and three north of Nipigon.

With the surveying of Airy Township, adjacent to Algonquin Park, in 1896, all of

A surveyor's field notes during the late 19th and early 20th centuries often combined fastidious drawings and paintings with the less romantic science of measurement.

southern Ontario and the accessible north (which is to say all territory south of the CPR line) had been subdivided into townships. More than a century had passed since Governor Haldimand had, in the interests of the arriving Loyalists, sent his deputy surveyor general, John Collins, from Quebec to Kingston to survey the first township in what was now a thriving province.

That the surveyed part of Ontario had taken more than a century to peg and parse, and that the surveyed portion was still just a fraction (less than a tenth) of the province's land mass, was routinely said by newspapermen, essayists and speechmakers to "beggar the imagination."

Fortunately, the imaginations of the audacious early members of the AOLS were far from beggared by whatever challenge, intellectual or geophysical, was posed by the mysteries and promise of the north. It would seem, rather, that the surveyors were *inspired* by the thought of the Land Beyond – of seeing it, measuring it, preparing it for what they perceived to be a profitable and productive future.

In 1896, the association, acting on the vision of its president, the redoubtable Willis Chipman, formed what it called the Committee on Exploratory Surveys. Under the chairmanship of Elihu Stewart, who had honed his instincts for the north surveying townships along the CPR line, the committee urged the government to undertake a systematic survey of the entire northern expanse of Ontario – from the CPR line north to James Bay, from Manitoba to the Quebec border. The politicians of the day took pleasure in

Elihu Stewart, seen at left, was a veteran northern railway and township surveyor who, as chairman of the Exploratory Surveys Committee of the AOLS, persuaded the Ontario government to go ahead with the Great Northern Survey in 1900. Right: The distinguished Haliburton surveyor, Alexander Niven, with his out-sized stern paddle and cedar-strip canoe, is seen here on the shores of Lake Timiskaming, circa 1896.

reporting that this immense tract of Ontario covered an area "the size of France and Spain combined" (one lecturer's estimate included Great Britain; another's, Poland). The survey, as Stewart proposed it to the government, would advance prospecting, mining, shipping, lumbering, roads – as well as the increasingly familiar possibility of (you guessed it) horses, harnesses and hay rakes. Apparently with farming on his brain, Stewart compared the province's current state of self-awareness – or the lack thereof – to "a farmer who, owning a 100-acre lot, has never considered it worth his while to examine the character of his possessions beyond the small clearing at his front door."

Because base lines and meridians were non-existent out there where the lines of longitude converge, Stewart floated the notion that the great survey could use the CPR track as its latitudinal foundation; it was well-staked, easy to locate, and, while not exactly straight, wasn't going anywhere soon.

At first Queen's Park resisted any commitment to such a survey – the cost, it was assumed, would be stratospheric. However, the Department of Crown Lands did decide in 1898 to run a meridian that would begin along the existing Algoma-Nipissing boundary and extend all the way north to James Bay. Wasting little time, Surveyor General Kirkpatrick hired the ever-ready and ever-capable Mr. Niven to get the job done. Which Niven did, accurately and quickly, even though he was now well into his sixties.

Predictably, Niven's report on the making of the meridian (soon to be named Niven's Meridian) praised much of the land through which the line passed for its *blazing potential as farm land* (anyone who knows the territory might be excused for wondering if, on their trips north, Niven and his crews were somehow tying into some magically hallucinogenic mushrooms or mosses).

At the annual meeting of the AOLS in 1899, Niven reported that of the 300 miles he had tramped on his way to James Bay, 200 traversed land "as smooth as the lawn in front of the Parliament Buildings." Given that the meeting just happened to be taking place at those same parliament buildings, the members needed only glance out the window at the bronze sculptures, the ornate lampposts, and the delicately landscaped grounds to get a convincing sense of what the north was all about. Niven further noted that the last 120 miles of his trek to salt water ran through a veritable agricultural paradise.

But still the government made no commitment to the great survey – until, unexpectedly, in 1899, the Liberal premier of the day, Arthur Hardy, resigned

David Thompson is often called history's greatest surveyor and geographer. But the artful and imaginative Peterborough surveyor, Sir Sandford Fleming, seen here in his sealskin coat, was not far behind. Like Thompson, Fleming was born in Scotland, and emigrated to Canada as a teenager. During the 60-odd years of his brilliant career, begun during the 1850s, he surveyed thousands of miles of rail track across Canada, created dozens of extraordinary maps (including "birdseye" maps in an age before aerial photography), co-founded the famed Royal Canadian Institute for surveyors and engineers, and was central to the invention of international standard time. While his plan for standard time was rejected by engineers and geographers at the International Meridian Conference in 1884, the plan (as represented on the map below) had by 1929 been accepted by all major countries around the world.

The dozens of men involved in the Great Survey of 1900 made their way into far northern Ontario first by train, then by canoe, and finally, if necessary, on foot.

and was succeeded by George W. Ross. Adrift in an apparent vacuum of ideas (at least of a sort that might catch the public's attention), Ross decided that the value of the proposed survey – the Great Survey of 1900 as it would come to be known – outweighed the fiscal cost. Quite suddenly during the first month of the first year of the new century, Ross gave the project his blessing, offering along with it a rather mingy budget of 40,000 dollars.

Under the guidance and fiscal whip of Surveyor General Kirkpatrick (who eventually brought the costs in at $40,518), the work was to proceed as soon as the snow melted.

And was to be finished by October!

It seems astonishing from the mesa of the present that Kirkpatrick expected the survey to be completed in a single season. Dreamer he may have been, but he understood, too, that The Great Survey would be one of the most difficult challenges in the history of surveying – and one of surveying's most prodigious accomplishments if it could be carried off.

In consultation with the ambitiously evolving AOLS, Kirkpatrick hastily laid down the battle plan. Rather than hire a single team of surveyors who would spread out across the province and proceed north like players in a game of Red Rover – or even hire two teams, or three, or four – Kirkpatrick hired no fewer than ten teams.

Each of them, headed by a surveyor, included a geologist, a timber assessor, a soil reader, and a biologist – as well as paddlers, provisioners, cooks, guides, and a crew of durable axemen. Each team would be responsible for assessing a portion of the north roughly the size of southern Ontario – which is to say some forty thousand square miles. Four of the areas to be explored were in the districts of Algoma and Nipissing (due north of lakes Ontario, Erie and Huron), four

A meal in the wilds – party #7, the Great Survey of 1900.

were in the district of Thunder Bay (north of Lake Superior), and two were in Rainy River district (far northwestern Ontario).

Acceding to experience, Kirkpatrick named AOLS war horses such as Thomas Speight and Hume Blake Proudfoot and the inimitable Mr. Niven, king of northern surveys, to head the teams. In the profession's unspoken hierarchy, Speight (whose name echoes today in one of the province's legendary surveying firms, Speight, van Nostrand & Gibson Ltd.) was viewed pretty much as Niven's equal, for his intelligence and sense of the planet. At the time, only a handful of surveyors – Willis Chipman, Elihu Stewart, Peter Gibson, Villiers Sankey – were held in the same high esteem. They were larger-than-life characters in a story about mythic lands, grand ambitions and epic travels. It was a tale replete not just with plights, plagues and uncertainty but with promise and what was perceived to be moral vigor.

Come spring that year, without fanfare, the surveyors assembled their teams and freight, and were away. Gone! First by train or steamer into the near north – then deep into a landscape that, except for the passage of fur traders during the 18th and 19th centuries, had for 10,000 years been populated exclusively by Algonquin, Ojibwa and Cree, or by their tribal forebears, and of course by eagles, muskellunge and moose.

The conditions of employment were clear: All travelling expenses were to be paid by the Department of Crown Lands, while each surveyor was expected to provide his own theodolite and compass, as well as a pocket sextant, micrometer telescope and plotting instruments. The government in turn supplied field books, a barometer, a camera, a metal case for preserving botanical specimens, as well as tents, canoes, cookware and food.

In the end, each team would be responsible for reports on geology, timber and flora; on fish, birds and mammals; on

Left: Dinner time on Quinze Lake – party #2, the Great Survey of 1900.
Right: Photo from an unidentified survey in northern Ontario, circa 1900 – captioned simply "Some Indians – Revillion Bros Post."

rivers and lakes, especially those that might suggest communication or transportation routes; and on "valuable water powers," including the volume and flow rates of rivers, and the nature of riverbanks, presumably for possible hydro-dam construction. It hardly needs saying that each team was also to report

on *the potential of the territory for agriculture* – even, it seemed, if there was no potential, or if they had to make it up, which to a degree some of them did.

Amidst the insects and privations, the long days of trekking, paddling, and hauling – all of this alongside the steady work of assessing and surveying and recording – the teams forged north, obliged to draw on every resource, every wile, every bit of persistence and toughness available to them if they were to complete their assignments. One group, Speight's, working in Nipissing District, travelled for five weeks, including two weeks by canoe, just to reach its point of commencement north of present-day Cochrane. The team's assignment, for starters, was to run a base line 72 miles eastward to the Quebec border from mile 198 on Niven's Meridian – and eventually to work north and west to Iroquois Falls, surveying as they went, finishing the season with a compass and micrometer survey of the Abitibi River and Lake Abitibi. It challenges credulity to consider that along the way the team consumed five tons – 10,000 pounds – of non-perishable food, caching much of it as the team moved deeper into the territory, so as not to starve on the way out. In addition, Speight's men ate several tons of fish, game and fresh berries.

Another group, led by Alexander Baird (who 25 years later would survey and drain the Holland Marsh north of Toronto), had its canoes severely battered and torn as it crossed Thunder Bay District on rivers that were too shallow and rocky to navigate. Over the course of weeks, the damage left the men's bedding and shoes and tents perpetually soaked,

ALEXANDER NIVEN • A RESOLUTE HAIRCUT AND A GAZE

At the time of his death in 1911, Alexander Niven, OLS nonpareil, was eulogized as the greatest surveyor Ontario had ever known. It is a claim that could still arguably be made, given that during a 45-year career Niven ranged into every extremity of the province, particularly the north, laying out townships and roads, defining provincial boundaries, and putting down base lines and meridians.

In 1896, under contract (as he frequently was) to the Ontario Department of Crown Lands, he completed his most famous and ambitious project, establishing a meridian that ran some 500 kilometres from Georgian Bay to James Bay and was subsequently christened Niven's Meridian. It was the first survey line in Ontario to touch salt water and is the line off which much of the north was subsequently surveyed and explored.

On that job, and many others, Niven encountered inestimable hardships, occasionally going hungry, sometimes existing with just an outdoor fire amidst snowdrifts and fierce cold. In February, 1884, after four days of paralyzing snow in northwestern Ontario, he wrote "I don't know how I can persuade the men to go another mile in such conditions. It is inhuman, but to stay put threatens our survival, especially with food running so low."

Born in 1836, in Niagara on the Lake, Niven attained his surveyor's license in 1859. The commonest surviving photo of him shows a stolid man with a resolute haircut and a gaze that surely echoed the one he employed to squint into blowing snow, confront bears (as he occasionally did), or assess the next patch of mosquito-plagued swamp. The photo is perhaps

Niven and his bride, Maggie McEvoy, photographed on their honeymoon in Montreal, in 1869, by the renowned photographer, William Notman.
(copyright McCord Museum I-39450.1)

equally a measure of the solemn patience with which he met his crews, no matter what their antics or complaints – and of his respectful efficiency in managing those crews. Month after month, year after year, he motivated his men to cut more line and explore more uncharted land than any other crews in the history of Ontario surveying.

Niven was similarly a manager of time – could get more out of a day than most people get out of two. His diary for 1869, a year in which he worked in Haliburton for the Canadian Land and Emigration Company, is a document of such persistent and enervating application, under frequently nasty conditions, the modern reader might well experience exhaustion just reading it. In entry after daily entry, Niven records that he snowshoed fifteen or more miles with a 50-pound pack on his back, slept in an unheated cabin, ate beans or nothing, melted snow for tea, surveyed such-and-such a tract of land, then another one ten miles away, then another, all the while looking for boundaries he could not find, listening for streams he could not hear (sometimes in the dark), keeping an eye out for wolves he could hear all too well. Often at night he would return a dozen or more miles to town for a few hours of office work by lamplight, then would head out again at 6:00 a.m.

Only once in his meticulous record of all this stupefying activity does Niven mention a few moments of ease, noting that late one night, by lamplight, in an unheated log cabin, he "began reading Dickens's David Copperfield" – a treat to be sure, at a time when the great Victorian novelist was still writing.

Present-day surveyor Curry

Bishop of Haliburton noted in February, 2015, that over the years he (Curry) has come across rotting cedar monuments bearing the initialed iron nails that Niven invariably and proudly drove into his wooden markers. "A lot of these posts would have been planted in snowstorms or amidst clouds of black flies," said Bishop, "and always they're exactly where they should be, exactly where Niven recorded their placement. He was as accurate using the crude equipment of his day as many modern surveyors are with GPS."

Beyond his field work, Niven was a thoughtful and persistent presence in the founding of the Association of Provincial Land Surveyors during the mid-1880s. He served as its president in 1888 and headed a variety of its committees before and after its incorporation as the AOLS in 1892.

Back home in Haliburton, where he was at best a part-time husband and father, he built the Presbyterian church, often preached in it, and served as reeve of the town for a dozen years – all such activity fractured by his annual departures into the wilderness to survey yet another hundred miles of the wildest terrain on earth.

To some, Niven's ardor for the rugged life might seem a form of masochism or slow suicide. But for Niven it was salvation. Even during his later years, when he was suffering from cancer, he would rally in the spring, hire a crew, head north, declaring on one occasion that this is how he chose to live, and perhaps to die, and that a few weeks in the wilderness went a long way to healing what ailed him.

Like many parents of his day, Niven experienced heartbreak over the deaths of children – in his case, two sons, the loss of whom catalyzed the breakup of his marriage to Maggie McEvoy, a marriage that surely knew more than its share of pressures. It is perhaps indicative of his extraordinary commitment to surveying, undoubtedly at the expense of family life, that no family members were present when he died in the Victoria Hospital in Toronto on May 7, 1911. Niven is buried in the cemetery in Haliburton, within a chain length or two of where he lived.

Husky sled dog pup, northern Ontario – Great Survey of 1900.

then rotting, and finally in tatters.

Elsewhere in the same district, Hume Proudfoot's team searched in futility for the waterway they had been told would lead them north to the Albany River. They were informed eventually by Aboriginal scouts and Hudson's Bay Company traders that such a route did not exist, that the charts they had been given by Kirkpatrick's people in Toronto were fiction. On hearing this, Proudfoot, having uttered a profanity or two, is said to have flung the charts unceremoniously (and much to his crew's delight) into Wabinosh Bay on the north corner of Lake Nipigon – and to have proceeded, this time more confidently, on directions from the local tribes.

Miraculously, none of the attendant hardships or frustrations prevented any of the teams from fulfilling its mandate – or at least from coming close. Back they all straggled in late September or early October, some having to crash their paddles through the skims of ice that had begun to narrow the rivers. They were broken, bruised, exhausted – out of food, out of footwear, out of patience. Two surveyors who had worked in Thunder Bay District, Walter Davidson of Lambton County and Joseph Tiernan from Tilbury, both in their mid-40s, all but crawled home, perhaps already suffering from the typhoid fever that would kill both of them within weeks of their return.

Meanwhile, the surveyors claimed their modest pay and, more importantly, at least from the government's perspective, handed in their reports: news of gold glimpsed here, silver there, copper over there; of timber to last for what was perceived to be an eternity. Such and such a river, they reported, was wide and open, its banks rich in iron, while another would be perfect for the shipping of cargo north to Hudson Bay. They returned with maps and plans and lists. And more stories than they could tell.

For the first time in the history of wilderness exploration, they also returned with photos – unreconstructed likenesses of how it was out there on the rivers and rock and in the forests. Those remarkable photos, some of which are shown on these pages, provide an inkling of the vastness, aloneness and grandeur of the north; and of the hominess of the survey camps: the men sipping tea on a Sunday afternoon, or smoking their pipes and chatting after dinner; or the men at work – chopping trees, portaging canoes, taking a bearing from a height of granite, all of it at a latitude where, in summer, the sunsets came late and the sunrises early.

Left: Sunday afternoon tea – Great Survey of 1900, party #1. Top centre: A cariboo luncheon – Great Survey of 1900, party #8.

Bottom centre: The bishop's cattle at Moose Factory – Great Survey of 1900, party #6. Right: Dog Lake traverse – Great Survey of 1900, party #8.

What the photos do not show is anything even remotely decipherable as farmland. And yet, predictably, the reports by Niven, Speight and several others trumpeted radiant possibilities for agriculture in what was now being referred to quite brazenly as "New Ontario." It was as if the "old" Ontario (yes, that one, the one that just yesterday was itself the glittering tomorrow) had been reduced to a relic, eclipsed by an alluring new neighbour to the north – by wealth, by mystery, by as-yet-untold potential.

And, indeed, the north was persuasive – was for some a kind of wilderness Canaan, a compelling new focus for the voracious spread of industry and colonization.

Anxious to make hay, literally and metaphorically, officials of the Department of Crown Lands were quick to knit up the ten teams' reports – inventing of them a document all but scandalous in its misrepresentations. The 25,000 square miles that comprised the Great Clay Belt were described as "equal in fertility to any land in older Ontario" and possessing "an equable and temperate climate." Elsewhere, the document boasted that the northern climate presented "no obstacle to successful agricultural settlement, completely dispelling the erroneous impression that the area's winters are of Arctic severity and the summers too short to enable crops to mature."

Such was the determination to lure aspiring farmers to the north that the dubious rewards of boreal agriculture were soon being promoted in a vigorous national marketing campaign.

In no time, surveyors were divvying the Clay Belt into townships, farm lots and roads. By an act of legislation in 1901, "northern farmland" was offered as a "reward" to Canadian soldiers who had served in the ongoing Boer War, as well as to the militiamen who had put down Riel in Manitoba or quelled the Fenian Raids in Ontario. The offer was extended not just to soldiers but to chaplains, stretcher carriers, nurses, and even to next of kin (essentially, to anybody with the drive to clear a few acres of forest).

Late in 1901, the Ontario government opened a land agency in London, England, offering "New Ontario farmland" to British immigrants. The result was that in 1902, 14,000 "location tickets" were awarded, each entitling a farmer (as well as many a non-farmer) to what a skeptical Toronto newspaperman referred to as "160 acres of misery." In some cases, ticket holders never showed up to claim their bit of Eden. Some appeared and, with commitment and luck

This 1907 photo of the survey training camp at Larder Lake, Ontario, includes at least two commissioned land surveyors: Homer Sutcliffe, OLS #388, standing directly behind the left tripod, and James Herbert Smith, OLS #345, holding the right tripod. Below: Surveyor Walter Davidson, OLS #201, of Lambton County, led a party into Thunder Bay District during the Great Survey of 1900. He returned frail and broken, and died within weeks from typhoid fever.

(and often large families to do the stupifying work that was required), made good on the area's ostensibly endless promise. Many more eked out a winter or two in poorly constructed cabins, sometimes shared with animals, and quit in despair. The Canadian historian Joseph Schull remarked in 1978 that "the north during the early 1900s seemed to be a sieve, emptying itself through abandoned farms as new farms were being established."

This remarkable cycle of false advertising, hope and abandonment, not to mention occasional success, continued not just for a year or two, as might have been expected, but for more than two decades. It was abetted by, among other developments, the coming of the Temiskaming and Northern Ontario Railroad into the territory in 1908, and eventually the Canadian National Railway, as well as initiatives in prospecting, claim-staking and mining, and of course in the timber and pulp industries.

In 1912, the province passed its Northern Development Act and committed five million dollars to northern industrialization. Part of that sum went to surveyors, but most of it went to a work force of 5,000 men brought into the territory to do the blasting, digging and gravel-hauling required to produce the territory's first network of decent roads.

The farm caper might not have gone on so long had it not been for the ascendency of William Hurst to the premier's office in 1914. Hurst had been Minister of Lands, Forests and Mines, and as premier was an all but rabid proponent of northern farming. One of his last gambits as leader of the province, in 1919, was to make a concerted appeal to Canadian veterans of World War I to buy farm land in the magnificent New Ontario for half a dollar an acre – or as the government's marketing posters put it, to embrace the "pot of gold at the end of the rainbow" and thereby participate in "the development of the finest farmlands in Canada."

Like the poster displayed on page 84, the one at right was produced circa 1918 by the Ontario Ministry of Lands, Forests and Mines, with the hope of attracting farmers – or at least dreamers – to the end of the rainbow in northern Ontario. Below: Timber scribes such as this were used by early surveyors to mark trees or to cut information into wooden boundary posts.

(Some 60 years hence, the Toronto land surveyor "Red" Petzold would comment that land "bought for fifty cents an acre in 1919 was still worth fifty cents an acre in 1980.")

What the campaign's posters failed to mention was that any farmland that could accurately be described as "the finest in Canada" lay not in New Ontario but 300 miles to the south, or a thousand miles to the west. Likewise, the posters failed to mention that aspiring northern farmers might better have beaten their plough shares into axes or picks, and to have staked their claim in timber, mining or hydro production – endeavors in which Ontario land surveyors were already profitably at work, hastily and skillfully preparing the way.

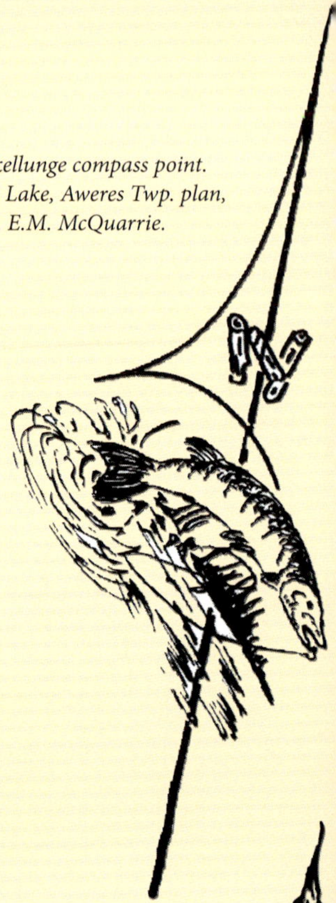

Muskellunge compass point. Maki Lake, Aweres Twp. plan, 1960. E.M. McQuarrie.

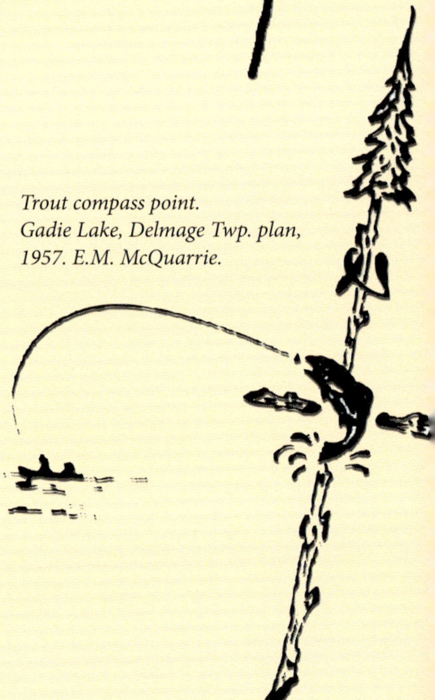

Trout compass point. Gadie Lake, Delmage Twp. plan, 1957. E.M. McQuarrie.

The field notes from Ontario surveys undertaken in the late 1800s and early 1900s were often adorned with meticulously crafted illustration, as exemplified here and on the opposite page in Willis Chipman's pictorials and graphics from his survey of Mulock Township in 1884 and of Lockhart Township in 1890.

Lake Huron traverse plan, St. Edmunds Twp., 1939. J.W. Tyrrell.

Bass compass point. Hurlburt Twp. plan, 1960. E.M. McQuarrie.

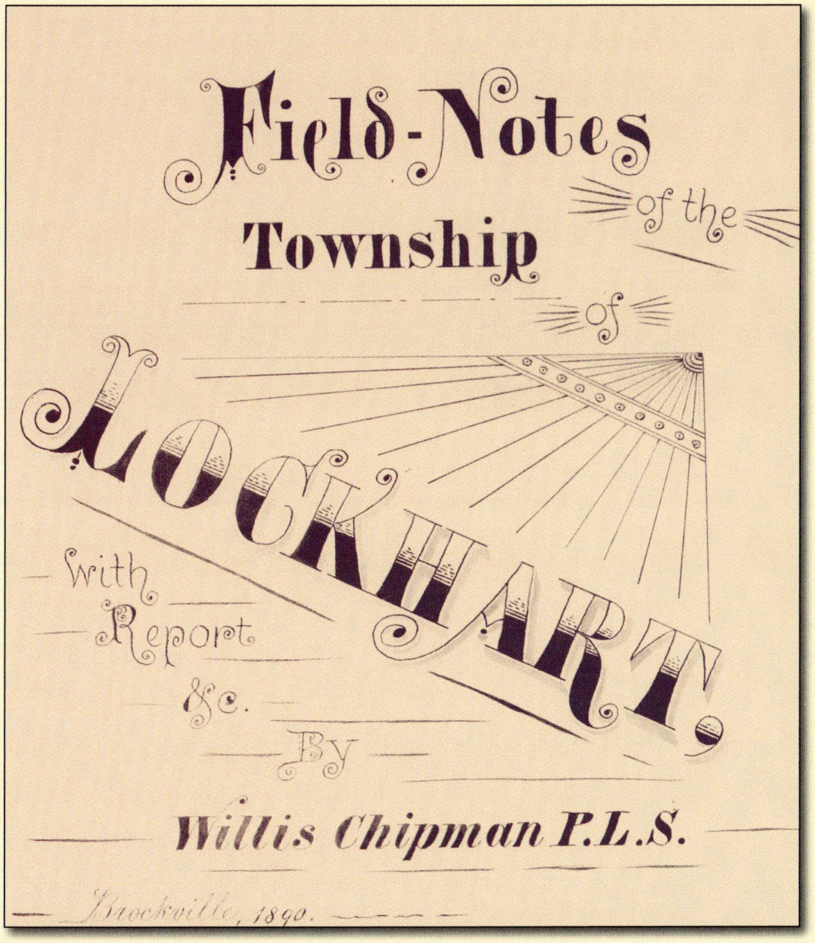
Field-Notes of the Township of LOCKHART, with Report &c. By Willis Chipman P.L.S.

Brockville, 1890.

Above: This elaborate compass point, drawn by Thomas Speight on his 1893 plan for Loughton Township, might at first suggest a citizen of ancient Rome. Closer examination will reveal that it is in fact Speight's fanciful interpretation of a First Nations warrior, with feathered headdress, tomahawk, bow, quiver, moccasins, and spear.

25

The Good Roads Train, *oil on canvas, by Moma Markovich, 1966. Government of Ontario Art Collection, Archives of Ontario 101257.*

CHAPTER THREE

HIGHWAY TO HEAVEN

FOR THE ONTARIO LAND SURVEYOR during the first decade of the 20th century, northern Ontario was both a riot of activity and a rolling bet on the future. And in due course we shall revisit those bracing latitudes.

Meanwhile, let us return to a more sedate part of the province where, by the miracle of the imagination, the year (as in the time of the Great Survey) is again 1900. The month is June. The place is a town in southwestern Ontario – perhaps your town. It is evening; the scent of lilacs is in the air. From somewhere in the distance to the east, a train's whistle carries across the rooftops. Women in ankle-length skirts and men in three-piece suits quicken their step toward the station.

All week, the lampposts and hoardings along Main Street have been papered as if the circus were about to appear. And in a sense it is – although not in any form that P.T. Barnum or Charles Ringling would have recognized. In the forthcoming version (which is just this moment crossing the municipal line), cars carrying elephants, giraffes and brightly painted wagons have been displaced by flat cars carrying long-necked steam shovels and deep-chested bulldozers, as well as some 800 barrels of Ontario cement. In the passenger coach, twenty or more well-practiced "performers" – surveyors, masons, carpenters – are on their feet, gathering their kit, preparing for arrival. They are Toronto men, recently hired, and they are under the command of the inestimable ringmaster, Archibald "Good Roads" Campbell.

Mr. Campbell, a land surveyor of the first order, is the province's newly appointed (some would say self-appointed) Commissioner of Highways. And as the train chuffs into the station, nearly 400 expectant witnesses have gathered by the tracks to see him, to hear him, perhaps even to shake his hand. They have been lured by posters proclaiming that the "Good Roads Train" is on its way, and that at the behest of Mr. "Good Roads," the province will build, free of charge, for any town on the growing Ontario rail system, five thousand two hundred and eighty feet – "an undisputed mile, ladies and gentlemen" – of hard-topped, deep-based, twentieth-century road! Will build it anywhere the town wants it to go.

Conversation on the streets has been lively since the posters appeared *(COME OUT AND MEET MR. CAMPBELL! WITNESS THE FUTURE FIRST HAND!)*. Some residents would like to see the area's first mile of hardtop built from the town square, out across the tracks, as far as the old Colonization Road. Others envision a mile of solid concrete where First Avenue is sinking slowly into the mud.

The Commissioner's ambitious road-promotion scheme – essentially, a publicity stunt – involves not just one but five "Good Roads Trains" in different parts of the province. The rolling stock is on loan from a trio of obliging railway companies – outfits that despite their commitment to rail can foresee the advantages of having their freight and passenger terminals connected to a web of highways. Financially, the project is backed by an Ottawa Valley newspaper magnate, who, like the railway owners, can see how a network of paved roads would significantly enrich his prospects.

In effect, the Good Roads Train is Campbell's grandiose way of spreading the news that Ontario needs highways. And every time a Good Roads Train pulls into a town to begin six weeks of dusty road-building, the well-travelled Mr. Campbell is aboard.

The train has barely stopped when a porter appears out of the recesses behind the coal car, leaps to the ground, and unfurls 40 feet of bright red carpet alongside the tracks. As the crowd presses forward, a natty little Punchinello with a head as square as a rail tie emerges onto the platform of his private Pullman car. It is Good Roads himself – waving, beneficent: a surveyor whose brassy and polished theodolite is focused aggressively on the future.

The town's mayor pushes forward to greet him. The crowd cheers. Rarely do they get to meet a government official, let alone one as famous and newsworthy as the ebullient Mr. Campbell. Just that afternoon, a photo of his Eminence at the wheel of a motorized vehicle dominated the front page of the local newspaper. More than fifty such vehicles are rumoured to exist in Ontario.

"Ladies and Gentlemen!" bawls Good Roads through a megaphone that has been handed up to him. "Thank you so *very* much for coming out to greet us! It is a momentous occasion!"

Within ten years, he tells the crowd, there will be four thousand motorized vehicles in Ontario!

Considering that by then there will also be some thirty thousand rail cars in the province, 4,000 is hardly a breathtaking number. However, for most of those present (few of whom have ever seen a motor car), the number is a head-shaker, nearly impossible to comprehend.

"I am talking not just about passenger automobiles," hollers Good Roads, "but about cargo-carrying trucks – trucks of the sort that, like box cars, like flat cars and tanker cars, will eventually carry commodities of every sort to every corner of this great province!"

Despite the skepticism, the echoing phrases bring another burst of cheering.

"The message I am here to impart," shouts Good Roads, "is that to accommodate this new river of traffic, we will need a first-rate system of paved roads! For the past thirty years, the focus of provincial transportation has been rail. And all power to our magnificent railways! Meanwhile, the age of the internal combustion engine, of rubber tires, of independent road travel is upon us!"

Another explosive cheer.

"What I am about to propose, in your gracious presence," intones Campbell, "is the inauguration of a provincial system of roads and highways that within fifty years will be unparalleled in the industrial world… And Mile One of that system will be surveyed right here in your fine town tomorrow!"

Some said it was the Scot in him – others, the trencherman or huckster. But more than any industrialist or politician, Archibald "Good Roads" Campbell is regarded as the leading force in converting mud trails such as the one pictured below to a paved highway system in Ontario. Campbell is seen here as a young surveyor in St. Thomas.

WILL IN FACT BE SURVEYED by the Commissioner himself.

More or less surveyed.

At 11 o'clock the following morning, Campbell applies his right eye briefly to the transit that has been set up in front of the Presbyterian Church and, having brought himself upright, signals to the pole man a hundred feet away and waves exuberantly to those who have gathered. And glances at his pocket watch.

Having acknowledged a last smattering of applause, Campbell doffs his fedora, is ushered to a waiting carriage, and is whisked away – off to the railway station; on to the next town, there to initiate another mile of hardtop.

And, thus, under similarly theatrical circumstances, do paved roads get their start in Sarnia, in Barrie, in Guelph… in Oshawa and Stoney Creek and Brantford.

NEARLY A CENTURY HENCE, Archibald Campbell would be identified as the man who, more than any other, was responsible for inspiring and building the provincial highway system.

By 1910, as Good Roads predicted,

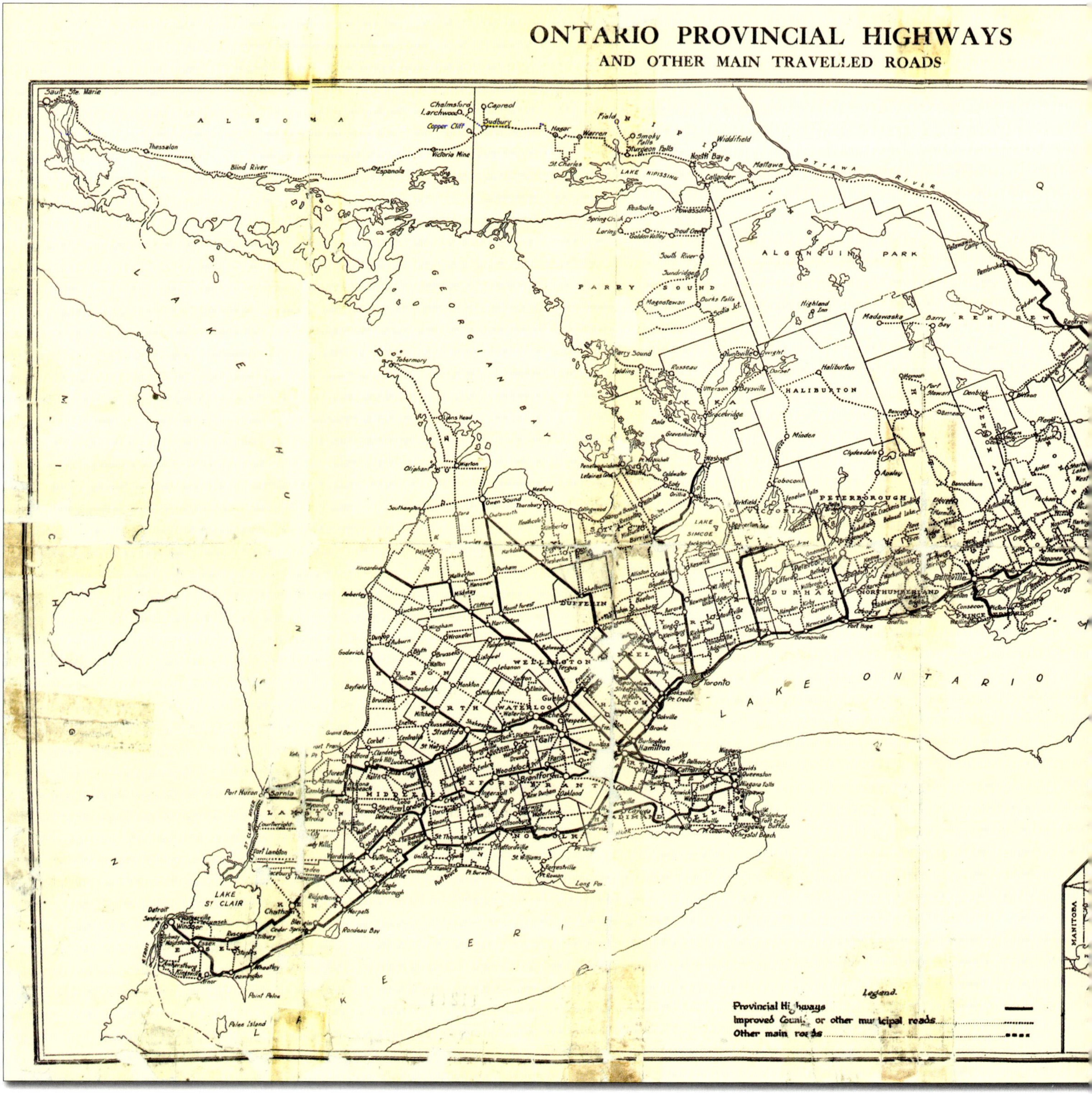
ONTARIO PROVINCIAL HIGHWAYS
AND OTHER MAIN TRAVELLED ROADS
Legend.
Provincial Highways
Improved County or other municipal roads
Other main roads
LAKE HURON
GEORGIAN BAY
LAKE ONTARIO
LAKE ERIE
LAKE ST CLAIR
ALGONQUIN PARK
Sault Ste. Marie
Sudbury
North Bay
Toronto
Hamilton
London
Windsor
Sarnia
Kingston
Owen Sound
Guelph
Peterborough
Belleville
Brantford
Stratford
Chatham
MANITOBA

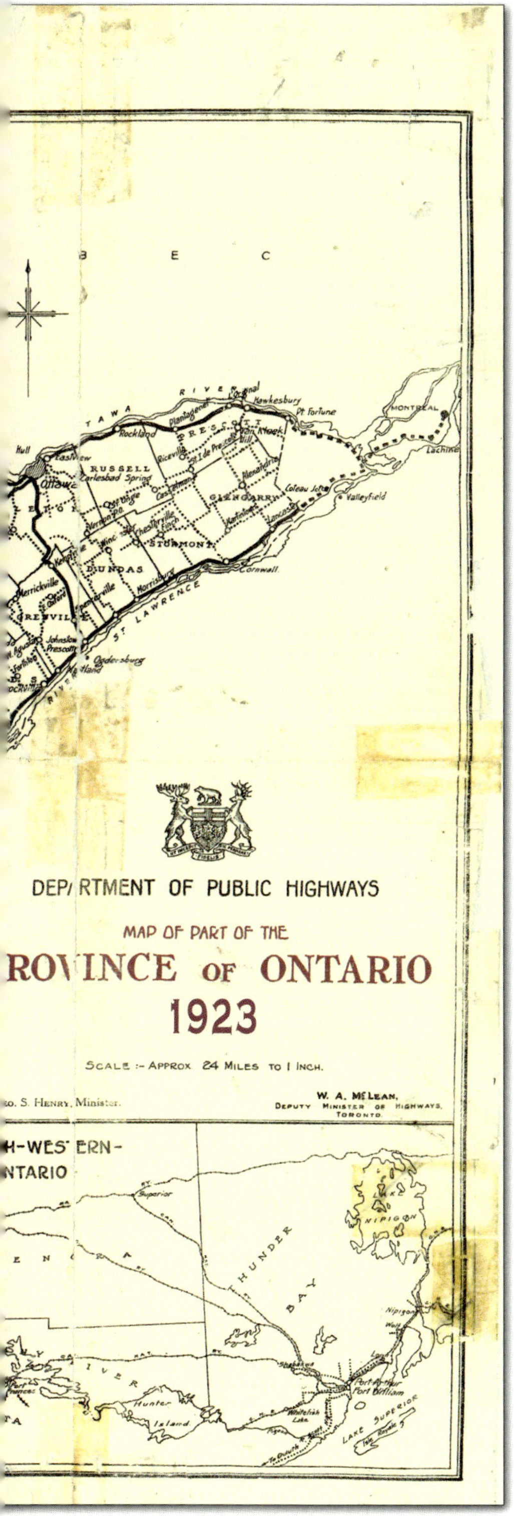

Left: This small map, published in 1923 (and printed here at about two-thirds its original size), was the first motorists' guide to Ontario highways. The map-cover at right, from the 1930s, is a nod in the direction of early women's liberation. It might suggest, less progressively, that women are more ardent than men in their requirement for clean toilets – and, by extension, more likely to be cleaning them.

there were indeed some 4,000 motor vehicles in Ontario – and by 1914, nearly ten times that number.

By 1922, despite the four-year interruption imposed by World War I, there were some 250,000 cars and trucks rolling on 3,000 kilometres of paved road in the province.

Five years later, nearly half a million residents of the province held Ontario driver's licenses obtained at the tolerable expense of one dollar – no test required.

In 1923, surveyors and cartographers teamed up to produce the first provincial road map – a map so quickly out of date, what with the proliferation of new highways, that Ontario's 663 service station operators were routinely asked by drivers to help them pencil route changes and short cuts onto their ever-evolving maps.

As much as any indicator in the province, the rapid expansion of highways between 1900 and the beginning of the Great War – and the subsequent expansion of those highways during the aptly-named Roaring Twenties — signaled boom years for Ontario surveyors. A simple correlation between the building of roads and good fortune for surveyors might be assumed: road-building required surveyors. But there was a more nuanced connection, in that new roads and a need for surveyors were tandem coefficients of expanding cities, expanding hydro production, the expansion of mines, lumbering and manufacturing; and of expanding agriculture as the need for, say, wheat and corn and animal protein brought a requirement not just for accurately measured fields and drainage but of course for more highways (which required more surveyors, etc.).

By the mid-1920s, the Department of Highways (which would eventually be described as a "factory" for the training of surveyors) employed nearly fifty

Far left: The famed surveyor and northern explorer, James W. Tyrrell, OLS #207, in the seal-skin parka he is reputed to have worn on his expedition to Hudson Bay in 1905. Close left: Irish-born Thomas Nepean Molesworth modelling the white woollen shawl coat that he wore while surveying the northeast shore of Lake Superior in 1856. Below: Surveyors in government-issue winter wear, as depicted by Moma Markovich in his painting, Winter Survey Party, *oil on canvas, 1962.*

Government of Ontario Art Collection, Archives of Ontario 101255.

SURVEYING IN STYLE

During more than 200 years of surveying in Canada, working garb has ranged from furs and skins to blanket coats, to khakis and fedoras, to the plaid shirts and blue jeans of the mid-late 20th century, to the hardhats and fluorescent vests of today. Some surveyors, such as those in the accompanying photos, have left their mark not only on the planet but on the world of fashion.

Above: During the late 1800s, surveyor Joe Cozens was an ambitious entrepreneur in Sault Ste. Marie, where he is said to have worked habitually in a tailored suit and wide-brimmed fedora. Right: The wealthy Ottawa-area surveyor, William Thistle, in his Persian lamb coat and matching ushanka, circa 1885.

Above: Charles Herbert Pinhey, OLS #250, sporting the magnificent double-breasted overcoat with mink collar that he wore on surveys in his adopted city of Ottawa.

Opposite page: Looking West on Highway 401 from Keele Street Viaduct – *an oil painting on canvas, by Moma Markovich, 1960. The same scene today, nearly 60 years later, is a variegated array of billboards, freeway signs, lighting poles, warehouses and condo towers – as well as 12 lanes of traffic. One thing that has not changed significantly since 1960 is the Ontario Government Downsview Complex on the right.*

Government of Ontario Art Collection, Archives of Ontario, 101220.

King George VI and Queen Elizabeth (the Queen Mother) are seen here (if barely) in 1939, in a parade car, arriving in St. Catharines to open Canada's first four-lane highway, the Queen Elizabeth Way.

licensed surveyors fulltime. And these were but a fraction of the self- or municipally employed surveyors engaged in laying out city streets, secondary farm roads and back highways.

By the mid-1930s, seemingly in defiance of Depession economics, construction had begun on the world's first significant "twinned" highway, the Queen Elizabeth Way, or QEW, between Toronto and Niagara Falls.

A mere decade after that, rumours began seeping from Queen's Park concerning a radical four-lane expressway that might eventually carry vehicles some 800 kilometres from the Ontario-Quebec border in the east to Windsor and Detroit in the west.

Given that there were just 600,000 motor vehicles in the province (compared to 13 million in 2015) and that a perfectly good railway and highway already existed along more or less the same route, the idea was perceived by most Ontario residents as a lurid excess, if not downright ridiculous. At Queen's Park, Frederick Robinson, an opposition MPP, pointed out that every car in the province could comfortably be parked on such a highway, and wondered facetiously if there were not "a more practical scheme on which the government could spend the hard-earned money of its citizens." For example, could a pipeline not be constructed to "carry maple syrup from Eastern Ontario to Toronto?" … or a "vast umbrella be raised during blizzards" to keep snow off the newly dubbed Golden Horseshoe of Industry? In the spirit of the protracted joke, a Toronto *Telegram* correspondent suggested the construction of a network of tiny gas pipes that would deliver a miniature flame to every Ontario residence, permitting citizens "to

light their pipes and cigarettes without the nagging inconvenience of striking a match or flipping a Zippo."

As contemporary Ontario residents know well, the spirit of Good Roads prevailed. Construction on the mega-highway began during the late 1940s, leading by 1960 to what we now know as Highway 401 or the MacDonald-Cartier Freeway.

The Commissioner would have smiled to know that, as of 2017, there are some 21,000 kilometres of paved highway in Ontario, and that in the Toronto area, where the 401 is twenty lanes wide, the vast expressway is the world's busiest thoroughfare, accommodating a half-million cars and trucks daily (this compared to the vaunted 405 Expressway in Los Angeles, the world's second busiest highway, which supports a mere 380,000 vehicles a day).

MEANWHILE, UP NORTH, surveying remained a mixed and expanding bag of opportunities: mining, claim-staking, timber surveys, hydro-electric development, the ongoing extension of the railways. Lumbering and the pulp and paper industry were by 1910 chewing up the evergreen forests with the persistence of long-horned beetles. Everywhere a pulp or lumber mill went up or a mine went down, a town was required, as were surveyors to lay out the streets, the sewers, the building lots, the banks. If government budgets ran dry, private wealth was there, spritzing money at whatever

The first wave of hopefuls in the Porcupine gold rush arrived in the north in 1909, travelling by train, then canoe. The women seen here may well have been prostitutes – "nurses" as they were sometimes called – who established camp next door to the makeshift saloon, in what is now Timmins.

facility or service was required to bring livability and a measure of zip to remote towns such as Cobalt and Cochrane and Timmins. At the time, the mines at Cobalt were turning out the present-day equivalent of two thousand dollars' worth of silver every minute, around the clock – about a billion dollars' worth a year. The Hollinger gold mine at Timmins was by 1912 producing 800 tons of ore a year, yielding a whopping quarter million ounces of elemental gold. The Porcupine gold rush as a whole – in Timmins, in Schumacher, in South Porcupine – produced more gold by far than either of the more famous gold stampedes in the Klondike and in California.

All of this is not to suggest that surveying in the north was becoming predictably secure or any less perilous than it had ever been. Mines were prone to fire and cave-ins. Towns were tinderboxes waiting to ignite. Boating safety was primitive. In 1905, Ontario surveyors were stunned to learn that Villiers Sankey, the chief architect of AOLS incorporation, and a respected head surveyor for the city of Toronto, had at the age of 52 drowned while surveying the route of the National Transcontinental Railway in the Rainy River District of northwestern Ontario. Sankey had been

JOSEPH BERNIER • ARCTIC DREAMS

When the renowned Canadian seafarer Joseph Bernier skippered the Coast Guard Ship *Arctic* far into the Northwest Passage in 1909, the voyage united Bernier's own dream of adventure and discovery with the dream of the visionary land surveyor Willis Chipman, who had long wanted to send Ontario surveyors into what is now the Canadian Arctic. Bernier was a rotund, rather unlikely-looking sea captain, in charge of an aging wooden vessel that was hardly the prototype of the durable ice-breaker. But the captain was a superior navigator, venerated by his crew; and the flexibility of the hull of his well-equipped vessel enabled it to endure the crush of ice that often surrounded it, where a hull made of the brittle steel of the era might have failed.

While colonial exploration of the Northwest Passage had to that point been undertaken exclusively by three nations, Norway, the U.S., and Great Britain, and while many of the *Arctic*'s features had been named for British royalty, Bernier's surveyors mapped the territory anew, and the captain (alas ignoring names by which the Inuit had referred to their islands and straits for centuries) applied "Canadian" names to the features of the land and water. More significantly, Bernier, in standard imperialist form, laid claim to the territory for Canada. In all, Bernier made ten trips into the Arctic (to go with his 270 Atlantic crossings) and in 1977 was honoured on a Canadian postage stamp. In Bernier's retirement, his beloved *Arctic*, moored at Lauzon, Quebec, was consumed gradually by local treasure hunters, or for firewood (its oak decks were said to have burned splendidly), so that, today, nothing remains of one of the most storied and significant sea vessels in Canadian history.

The Arctic, *at Pond Inlet, 1923.*

Canadian Surveyors Take a Licking

Since 1908, when the Canadian post office issued a pair of stamps celebrating the country's first recognized surveyor and mapmaker, Samuel de Champlain, a total of thirteen Canadian postage stamps have honoured surveyors or surveying. Perhaps the most obscure of those stamps was issued in 1974 when Canada Post honoured William Hamilton Merritt, who trained as a field surveyor in Burlington during the late 1700s, and during the 1820s was instrumental in the building of the Welland Canal. A 1976 Canadian stamp, marking the American bicentennial, bore a likeness of Ben Franklin, a land surveyor and Founding Father of the United States. Other commemoratives have celebrated the work of David Thompson, sometimes called the greatest geographer and mapmaker in history; Joseph Bernier, who in 1904 took the first Canadian surveyors (all of them members of the AOLS) into the Northwest Passage aboard his wooden vessel, Arctic; and Sir Sandford Fleming, who during the 1880s surveyed much of what is now the CNR's trans-Canada route, and a few years later made a decisive contribution to the creation of global standard time. Early drawings for a 1961 stamp celebrating northern development featured a surveyor and his theodolite (below left), but by the time the stamp appeared, an alternate design had been chosen (right center), reducing the surveyor to a tiny image dwarfed by a massive piece of earth-moving equipment.

Fleming, it should be noted, designed the first Canadian postage stamp, the three-penny beaver, in 1851.

Left: Note the candles fastened precariously to the felt hats of these silver miners at Cobalt, circa 1912. Above: Smelter pouring molten gold extracted from the Hollinger mine near Timmins, which at the time (1913) was one of the richest gold mines in history. Below: The Porcupine Fire in 1911 occurred during a heat wave that pushed temperatures to 107 degrees Farenheit. Of the reported 200 dead, many were buried hastily in burlap shrouds, as seen here, while others were never found.

travelling on Lake Manitou with three assistants when a gale blew up and capsized their canoe. The lone survivor of the calamity described Sankey clinging to the canoe for an hour or more after two of his assistants had perished, and finally slipping beneath the waves. Sankey's remains were located several days later in 75 feet of water, just 50 yards from shore.

Men with theodolites and chains were doing an equivalent but less perilous business in the province's rapidly expanding towns and cities. By 1911, nearly 55 percent of Ontario's 2.5 million inhabitants were urbanites, and the construction industry was under pressure to provide homes and offices and schools, and of course infrastructure, to accommodate them. In Toronto, the suburbs were spreading outward to Swansea, Riverdale and the eastern Beaches. For lack of land and elbowroom,

The 15-storey Canadian Pacific Building, constructed in 1913, in Toronto, was at the time the tallest building in the British Commonwealth and one of the tallest in the world. In 1929, its terracotta cladding, which was deteriorating in the Canadian winters, was replaced with Indiana limestone. Opposite page: Until the construction of the CP Building, the Independent Order of Foresters Building, constructed in 1896 at Richmond and Yonge streets, was Toronto's tallest. As of 2017, both buildings are still standing. Near right: At the 1913 AGM, AOLS President, Thomas Speight, portrayed here, did his best to reassure the province's surveyors that the development of downtown Toronto (which, by comparison, must have seemed Trumpian in excess) would at least provide some of them with work.

Toronto's core was pushing skywards. The city's tallest building from 1896 to 1912 had been the 12-storey Temple Building at Richmond and Bay streets. This splendidly ornate wedding cake housed the world headquarters of the International Order of Foresters. But in 1913, as the Great War loomed, the Canadian Pacific Railway eclipsed the Foresters by constructing a 15-storey tower at Yonge and King streets – a building that at the time was the tallest in the British Commonwealth.

At the annual meeting of the AOLS, in 1913, Thomas Speight, now president of the association, informed a skeptical membership that the Royal Bank of Canada was about to construct a building 20 storeys (some 250 feet) high, kitty-corner to the aforementioned CP Rail Building. Famous skyscrapers of the future, such as the Canadian Imperial Bank of Commerce Building (1930) at King and Bay and the Empire State Building (1931) in New York, were as yet unimaginable, and consensus among the surveyors in Speight's audience was that big business had finally gone too far, that the

whole crazy competition in tower construction was, despite its opportunities for surveyors, messing with the contours of the city, disfiguring its beauty.

Speight appeased the membership by revealing that the demand for surveying in Ontario in 1912 had surpassed all previous records, thanks to rail- and highway- and city-building; and of course to the aforementioned developments in New Ontario, which Speight knew well, having been a central participant in the Great Survey of 1900. In the "glorious north," as he depicted it, minerals and precious metals were all but leaping out of the granite. Likewise, spruce trees were rocketing from the soil into the pulp and paper mills – and from there onto the doorsteps of Ontario, in the form of sensationalistic newspapers or, increasingly, glossy magazines about the British royal family and the stuffed shirts of the Prussian monarchy. As Speight saw it, the prospects for our old friend northern agriculture (which in reality were about as rank as a chainman's work socks) were on the brink of achieving a new high. With the increase in opportunities, he told the membership, there were now, on average, seventeen surveyors a year entering the profession in Ontario – enough to sustain the ranks into the thick of the burgeoning prosperity.

Little did he suspect that under the pall of the coming wars – and, more so, of the Great Depression – the influx of surveyors to the ranks would shrivel almost to nothing and would not hit current highs again until the mid-century boom, some four decades later. During the 1930s, there would in fact be a shrinkage in membership, as once-affluent surveyors cast about for work, unable to pay their dues.

Meanwhile, the explosion of the urban population stimulated an unprecedented demand for food, and thus an immediate need for more farms, and more productive farms. During the summer of 1909, surveyors in southwestern Ontario laid plans for, and oversaw, the installation of 30 million linear feet of drainage tile, significantly improving some 200,000 acres of soil.

At Holland Marsh, near the tip of Cook's Bay, on the south shore of Lake Simcoe, one of the most famous and successful drainage projects in Canadian history began to take shape. Bisected by the Holland River (named for Upper Canada's first surveyor general, Samuel Holland), the marsh had long been considered a potentially valuable agricultural resource. For centuries, the local Huron and their predecessors had harvested it for edible plants, waterfowl,

A surveyor on the east bank of the Don Valley assesses the construction of the Bloor Street Viaduct, December 2, 1915. At the time, shortsighted Torontonians called the viaduct "the bridge to nowhere," because at the time there was little in the way of development east of the valley. With the viaduct's completion, development of course thrived.

Above: The 200-foot steel tape was standard equipment among surveyors until the advent of tellurometers, total stations and GPS. This particular tape was manufactured in Windsor, Ontario, by the Lufkin Company, circa 1920.

muskrat, bullfrogs, and the like. For a decade at the beginning of the 20th century, the nearby Bradford Mattress Factory had sacked it for swamp grasses, tons of them, that the employees dried and bundled into canvas ticks, a primitive form of mattress.

In 1905, physicist William Day of the Ontario Agricultural College at Guelph began tests on the nutrient-laden marsh muck and determined that, if drained, the organic goo would be an out-and-out "supersoil" for the growth of carrots, onions, lettuce – in effect a kind of money salad.

But seed funding was slow to accumulate. Then came World War I, closing the financial sluices entirely until the early 1920s, when the provincial Department of Agriculture hired Alexander Baird, a Leamington surveyor, to draw up the drainage plans. Like Thomas Speight, Baird had successfully led one of the teams in the Great Survey of 1900. More pertinently, he had done extensive drainage work in the marshes at Point Pelee and in nearby Essex County, on the shores of Lake Erie.

Baird's plan called for the creation of 28 kilometres of canal, through which the Holland River would be diverted around the periphery of the wetland. Another 28

Opposite page: The Holland Marsh, with its rows of vegetables, as seen during the summer of 2016, and the rather eccentric-looking Samuel Holland, Upper Canada's first Surveyor General. Holland's name graces not just the marsh but the towns, Holland Landing and Holland Centre, as well as the Holland River, Holland Township, Holland Lake and numerous Holland streets and avenues in cities such as Ottawa, Toronto and London. Below: Surveyor Alexander Baird, a drainage specialist from Leamington, who led one of the parties in the Great Survey of 1900, was hired in 1920 to design the canals by which the Holland River would be rerouted and the Holland Marsh drained.

kilometres of earthen dikes would keep the water out once it had been pumped from the marsh.

To this day, the 21,000-acre market garden – referred to recently in the *Globe and Mail* as Ontario's "billion-dollar salad bowl" – produces a significant volume of the fresh vegetables consumed in south-central Ontario.

For all their productivity, however, the marshes are a hard-learned and anachronistic lesson in the culture's changing standards for land use and ecological integrity. In an interview in August, 2015, surveyor Ron Stewart of Marshall Macklin Monahan noted that a project such as the drainage of the marsh could only have been undertaken in the days before wetlands became legally protected ecological treasures and that, even the now-indispensible 400 Highway, built during the 1950s across the drained marsh, probably would not have been approved in its current location under today's more rigorous environmental laws.

AGAINST THIS BACKDROP of bristling rural and urban transformation, the maps representing the province's cities, farmland and forests remained woefully inadequate and fragmented during the early years of the 20th century. Nearly a century had passed since the matchless surveyor and explorer David Thompson – the "Epic Wanderer" as journalist D'Arcy Jenish called him – had mapped the territory around Lake Superior and from there west to the edge of the continent. By the standards of the day (and of one-man mapmaking), Thompson had created a magnificent counterpane, eight feet by ten, which brought together an intricate, sometimes byzantine, ecology of mountains, valleys and watersheds. When in 1814 Thompson sent his just-completed map by canoe from Montreal to Fort William (then the field headquarters of the North West Company of fur traders), the chief partner in the company, Simon McGillivray, along with other partners and associates, could not

When in the spring of 1814 David Thompson completed his map of the west, a portion of which is seen here, he is said to have had no idea that the grand document (as big as the topsail on a schooner and representing thousands of square miles of North America) was perhaps the greatest piece of map-making in history. The map's immediate fate upon completion was a precarious thousand-mile canoe ride up the St. Lawrence River and Great Lakes from Montreal to Fort William, the field headquarters of the North West Company of fur traders.

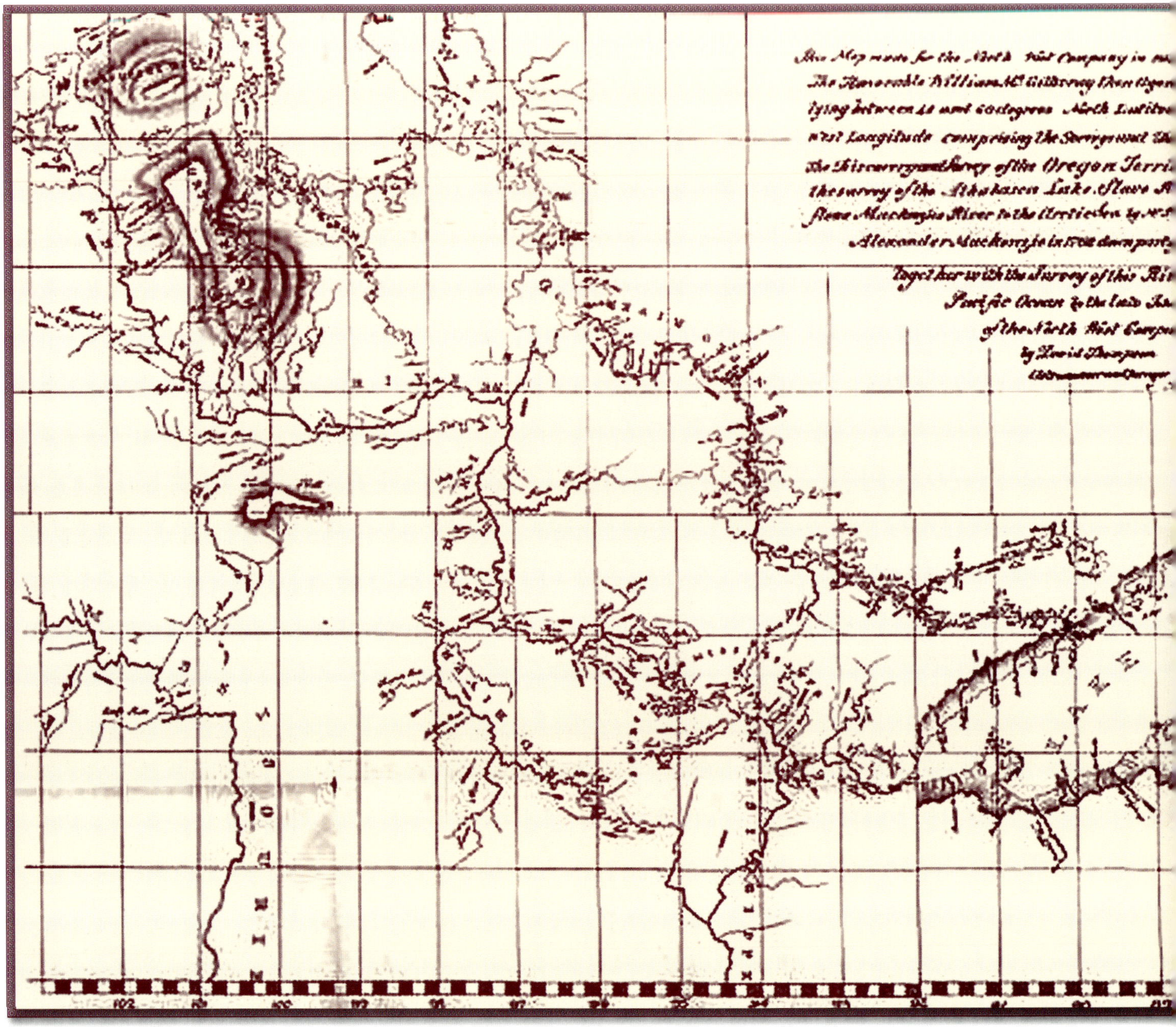

C.W. Jefferys' drawing shows David Thompson at work with his sextant and a refracting prism, undoubtedly of his own design. The Nicol prism, often used with a sextant, was not invented until 1828, some 15 years after Thompson's mapmaking ended.

stop staring at and marvelling over the thing, which at the time was perhaps the greatest piece of mapmaking in existence.

Thompson, it must be remembered, had amassed the components of the map using a nineteenth-century sextant and compass, and had organized those components working free-hand, in pencil, on notebook pages, guided as much by instinct and imagination as by the stars. The inevitable result was that, in its more remote corners and tracts, the great map, for all its splendour and shapeliness, lacked the accuracy and detail that 20th-century geographers required. And yet this hand-drawn, hand-lettered map (reduced to the size of a Kleenex tissue so that it could be included in Ontario's grade-six geography text) was, during the early 1900s, still the standard by which school children in the province were introduced to western Canada and, more specifically, to northwestern Ontario, a region roughly the size of France.

Professional geographers and engineers were hardly better off than the school kids when it came to maps of Canada and Ontario. Numerous district and county maps existed, as well as fragmented mining and geological maps, but in no way did these accurately add up to a likeness of the country or province as a whole.

As recently as 1904, Albert Montand, the chief geographer in the national Department of the Interior, had reported to those who demanded new and better maps that he was unable to produce such cartographic updates because existing maps and plans, from which new maps might be drawn, were so carelessly slapped together. As a control, he explained, he had conducted a number of surveys in Ontario, some of which showed that roads, rivers and township lines were, in some cases, up to two miles off-kilter. Surveyor and AOLS executive Willis Chipman pointed out that on a recent map produced by the Ontario Department of Crown Lands, dozens of lakes and rivers were plain missing, and that rivers were sometimes shown draining in the wrong direction.

What was needed to correct the situation was a coast-to-coast geodetic survey – a survey that would create a basic geographic reference system, one that would account for the curvature and dynamics of the earth and would serve as the foundation for Canadian maps as far into the future as anybody could see.

The AOLS had been agitating for such a survey for half a decade, and recently had sponsored what it called its Committee on Topographic Surveys. The group

Opposite page: Like dozens of others, this magnificently spindly wooden observation tower was constructed in eastern Ontario during the years before World War I for use in the Canadian Geodetic Survey. When the war interrupted proceedings, there was fear that the wooden towers would be rotten and unsafe by the time the survey resumed. And in large part they were, leading to the invention of transportable steel towers.

was chaired by Otto Klotz, who eventually would be known as "the father of Canadian astronomy." On at least one occasion, Klotz was also called "the most distinguished member of the Association of Ontario Land Surveyors" – no small recognition in an organization that included the likes of Chipman, Speight and "Good Roads" Campbell.

Under pressure from Klotz and his committee, the federal government was at last convinced, in 1905, that a national geodetic survey should proceed immediately (not a moment to lose after the government of Sir Wilfrid Laurier had sandbagged the idea for half a decade). That summer, the first triangulation towers, spindly wooden things, were built in the Ottawa area. More were added in subsequent summers, until by 1910, some sixty of these long-legged shacks existed across what skeptics now referred to as "Old Ontario." Surveyors manned the towers like flag-pole sitters, accomplishing daytime triangulation with the use of mirrors to reflect sunlight, and working the night hours using acetylene lamps that, backed by reflectors, could be seen for miles.

As the process trundled forward and the first accurate maps (initially from the Ottawa area) inched into being, the Department of Militia and Defence began using the geodetic survey as a guide to creating its own detailed maps, the first accurate military maps in Canada.

But all of this crashed and crashed hard on August 4, 1914, when Great Britain declared war on Germany, and Canada, as Britain's colony, was involuntarily (though willingly) brought to the fight.

Surveying and the Art of Ontario

Surveying has been described as the science of measurement combined with the art of law – which tends to suggest a rather tightly calibrated profession perhaps lacking the human touch. By contrast, surveying has also been a well-visited and less-rigorously defined subject for Ontario's visual artists. Since the early 19th century, our painters and sculptors have been attracted, for example, to the country's first recognized surveyor, Samuel de Champlain – and to numerous other surveyors who predate photography by as much as a hundred years. Fittingly, in a book that spans two-and-a-half centuries, the small gallery collected on these pages includes work from the 19th century (including our glorious cover image by the Toronto dream-realist, Paul Kane), as well as contemporary pieces, such as *Batchawana Bay*, by the Ontario surveyor, lawyer and watercolourist, Izaak de Rijcke. The styles of the collection range from de Rijcke's interpretive rendering of boundaries and landscapes familiar to the Ontario surveyor, to James Cook's folk depiction of his late father, the Sudbury surveyor Bev Cook, to works by the prolific 1960s documentarian, Moma Markovich. H.H. Leering's dazzling piece of mid-century realism, *Survey Crew*, appears on page 142.

Above: Newmarket surveyor Robert Garden's 1991 watercolour, Champlain with Astrolabe, 1613.

Left: Batchawana Bay, *watercolour on paper, by Guelph surveyor and lawyer, Izaak de Rijcke, OLS #1458.*

Right: Survey Party, *oil on board by Moma Markovich, 1961 Government of Ontario Art Collection, Archives of Ontario 101254.*

Below: Cover illustration from field notes, survey of Eton Township, Rainy River District, 1896, by Thomas Speight, OLS #185.

Lower right: Commemorative portrait of Bev Cook, OLS#1149, painted in 2013, by Bev's son, James Cook, (note the second image of Bev, with his tripod, by the road).

The Battle of Vimy Ridge, *oil on canvas, by Richard Jack, circa 1918.*
Canadian War Museum.

CHAPTER FOUR

WAR–VIMY RIDGE

A VICTORY FOR ONTARIO SURVEYORS

By *mid-autumn of 1914*, almost every Ontario surveyor young enough and fit enough to enlist – 45 of them in all – had laid aside his transit and chain, picked up his marching orders, and headed overseas. For the surveyors, the gun was ultimately a less serviceable battle tool than was the knowledge of measurement and angles and an understanding of topographical detail. No sooner were the young surveyors on British or European soil than many were put to work on the vital mapping of the terrain on which the war was taking place. Others plotted attack routes, or trajectories for cannon and mortar fire.

No story of Canadian surveyors in the Great War could possibly be told without noting the crucial work of Major Andrew McNaughton, an electrical engineer, originally from Saskatchewan, who, backed by a coterie of Canadian surveyors and technicians, pioneered the ingenious use of what he called "sound ranging," a method for locating the precise position of enemy artillery. The basis of the technique was the placement of four well-separated microphones, in a row, as much as a mile out in front of the Canadian battle line and parallel to that line. Further out than the microphones, a "listener" was stationed to record the moment at which the sound of an enemy artillery cannon reached him. Listeners linked to each of the four microphones would get signals from the front listener when a shell was fired and would record the time it took for the sound of the explosion to reach his microphone. By combining the auditory data with the known distances between microphones, and whatever angles were pertinent, men in command posts at the lines could triangulate exactly the point on their maps where an enemy gun was located. Often within a minute, the allied artillery was able to destroy such a gun with artillery fire of its own.

In darkness, a visual method called "flash spotting" was used for locating enemy artillery. Watchers at two well-separated posts along the allied line

The rather stern-looking Saskatchewan electrical engineer, Major Andrew McNaughton, directed dozens of Canadian surveyors and technicians in innovative sound-ranging and flash-spotting operations during World War I.

Below: Canadian ground forces emerging from their trenches on April 9, 1917, to begin their assault on the German-held Vimy Ridge.

Opposite page: Canadian surveyors mapping in France during World War I.

would look for flashes from enemy artillery and when they saw such flashes would press their respective signal keys, record the precise direction from which the flash came, and send an impulse to a command post, where surveyors used the timing and angles to triangulate the precise position of the gun that had been fired.

The techniques are said to have contributed significantly to the allied victory at Vimy Ridge, in April, 1917 – indeed are thought to have been decisive in turning the tide in favour of the 20,000 Canadians who fought the battle (a battle in which 3,600 Canadians died, and 7,000 more were seriously wounded).

Such techniques were not the only contribution of Canadian surveyors at Vimy, or at, say, the battle of Valenciennes in November of 1918. Only after the war ended could it be revealed that Canadian surveyors at Vimy Ridge had, in the days before the assault, used aerial photos and maps to reproduce, in an area behind the front lines, a full-scale simulation of the complex German defences atop the ridge. On that schematic stage, the Canadian troops went over and over the logistics of their assault, rehearsed and memorized every facet of it. When they poured over the Arras Escarpment at dawn on April 9, they were thus aware of every enemy strongpoint and the location of every enemy trench, and each man was able to carry out his assignment with directness and precision.

On the day before the armistice, McNaughton was rewarded for his efforts with a promotion to Brigadier General and an appointment as General Officer in charge of Canadian Heavy Artillery, both of which titles he carried into World War II.

In an address to the Association of Dominion Land Surveyors in 1926, seven years after the armistice, McNaughton noted that during the last hundred days of the war, some 73,000 tons of artillery shells had been fired by the Canadian corps, and that nearly every shell had

been aimed according to information from maps or data supplied by Canadian survey units.

In particular, he praised the surveyors for their persistence, accuracy and courage under fire, which he said had helped bring an early end to the war and had thus saved many thousands of lives on both sides – a vivid and defining tribute to AOLS members who were there.

Of the roughly 65,000 Canadians killed in the war, five were Ontario Land Surveyors. There are no statistics on how many of the 45 Association members who served were injured or suffered the catastrophic effects of, say, mustard gas or chlorine, but there are said to have been a dozen or more. Nor is it known how many of the 40-odd students who were articling in Ontario at the time became casualties. Because of their age, it can be assumed that many of them put on a uniform.

At the association's annual meeting in 1915, president Edward Wilkie read a letter written by an articling student named Kenneth Campbell, who at the time was serving in the Canadian artillery in France, undoubtedly under the command of Major McNaughton. The letter had been written to another articling surveyor back in Ontario, and said among other things:

> *... Please jog Mr. Wilkie's memory about my OLS exams. I think under the circumstances the association might let the odd month slide and let me write my finals when I get back. I'm not over here on any giddy pleasure trip.*

The tragic postscript to the story was that Kenneth Campbell, the letter's author, was killed in action several months later, doing the work that surveyors had been doing all along, and thus never fulfilled his ambition to return to Ontario, write his exams, and turn his surveying skills to peace-time activity.

Top to bottom: Sergeant Major Clarence Monteith of Stratford, Thomas Code of Cobalt, and Andrew Gray of Brockville were among a hundred or more young men who were either commissioned land surveyors when they enlisted to fight in World War I or became surveyors upon their return to Canada. Of the three, only Code was a surveyor when he signed up. For years after the war, he was part of a unique family record: five brothers registered as Ontario Land Surveyors at one time. Two members of the family's next generation, Robert Code and Bryce Code, also earned surveying commissions during the 1940s and '50s.

THE FOLLOWING YEAR, at the annual meeting of the association, in Toronto, Willis Chipman organized the first-ever Veterans' Dinner, at the Engineers' Club, for those with at least 30 years of experience as surveyors. A group photo from the event shows a rather dour-looking group of mostly white-haired, mostly mustachioed AOLS old-timers, who judging by the wartime frowns on their faces might easily have passed for a convention of undertakers.

IT WOULD BE A PREPOSTEROUS UNDERSTATEMENT to say that the Great War was hard on Canadians – killing 65,000 of them, wounding 250,000 more. Worldwide, the war left 17 million dead, 20 million wounded. If one tiny ray of utility found a peep hole through the mustard gas and trench dust and made its way to Canada, it was that the war, once ended, provided Canada with some 3,000 trained pilots, courtesy of the British Royal Flying Corps – this at a time when aerial photography was about to emerge as a force in Canadian surveying and mapmaking. The deathly back-story to this surfeit of pilots is of course that some 1,600 Canadian flyers, many of them in their early or mid-twenties, were killed in action, mostly during the latter years of the war, as flight took on a greater role in the fighting.

Nor did aerial photography as it pertained to surveying flourish immediately in Ontario after the armistice. For one thing, there was, until 1922, only one proper takeoff and landing field in the province, at the military base at Camp Borden, west of Barrie.

It was on that airstrip on a morning in early May, 1921, that two Canadian pilots, Captain Hubert Holland and Captain Edward Owen, both veterans of the war, zippered themselves into well-insulated flight suits with fur collars, donned leather helmets and goggles, and took off in an open-cockpit biplane, a deHavilland DH4, built for bombing raids over Germany. Their mission was central to a unique experiment: With the Eastman K-1 camera and the 25 rolls of film that were aboard, they were to fly southwest some 170 miles and take hundreds of aerial photos of the flourishing city of London, Ontario – the first time this sort of documentation had been attempted in Canada.

As the DH4 approached the city, Holland, who was at the controls, dropped the plane to 500 feet, located a scattering of men waiting by a truck on a rural road, and brought the aircraft in for a landing. Having picked up the details of their assignment, the flyers again took to the air.

During the next three hours, Holland and Owen guided the DH4 back and forth dozens of times in precise lines over London – 168 miles in all, at an altitude

In May 1921, Captain Hubert Holland and photographer A.E. Church, seen here (left and right) beside their deHavilland DH4 bomber, participated in the first attempt to create a vast aerial photo-mosaic of a Canadian city – London, Ontario. On a subsequent and related mission in April 1922, Holland died when the DH4 crashed. Below: The project's mastermind, Douglas Nelles, an RAF pilot, survived two air crashes during World War I.

of 5,000 feet. On each pass, they snapped off dozens of images, each of which would contribute to an enormous photo-mosaic, representing, in total, 42 square miles of the city below.

During the late afternoon, the trio, weary from the vibrations and noise of the plane, dropped off the film, arced the biplane into the northeast, and winged their way back to Base Borden. Like most Canadian flyers of the era, they relied for navigation on a simple compass, an altimeter, and on what they could see of the farms, forests and rivers below.

This milestone project was the inspiration of Simcoe-born surveyor, Douglas H. Nelles, a 30-year-old RAF pilot who had survived a pair of air crashes during the war and who believed aerial photography could be a revolutionary surveying and mapping tool, one that would bring about a radical reinterpretation of the planet.

In order to tie each individual photo into the mosaic, in perfect alignment and proportion, Nelles, a participant in the ongoing Canadian Geodetic Survey, established control monuments on the ground. Some were merely the usual brass pucks, atop posts and anchors, while others were as prominent as church steeples or office buildings, visible for miles around. The aerial work, it should be noted, was merely an adjunct to the aforementioned geodetic survey. Indeed, the London portion of that sur-

In 1915, as World War I consumed Europe, Hamilton McCarthy's bronze statue of Canada's original surveyor, Samuel de Champlain, was unveiled at Nepean Point, overlooking the Ottawa River, in Ottawa. Alert surveyors were quick to notice that the courtly looking explorer was holding his astrolabe upside down.

vey was the true wunderkind here, in that it took six years to complete, contained detailed drawings of every street, walkway and building footprint in the city, and ultimately occupied 61 large-format books of 200 pages each – a total of 12,200 pages.

Not surprisingly, it was the grand photographic mosaic rather than the 12,200 pages of exquisite drafting that caught people's attention (the aerial mosaic exists to this day, as big as a foresail, in the possession of the London Museum of Art).

Unfortunately, when Nelles and his team began piecing together the individual photos, they discovered a number of gaps in the assemblage (perhaps the camera had misfired, or territory had gone unrecorded while film was being reloaded). The result was that a few months later, in the spring, Holland and Owen were back in their DH4 en route to London to finish the assignment. Somewhere near Brantford, having encountered harsh weather and perhaps off course, they lost control of their airplane, and crashed, killing Holland, seriously injuring Owen. Beyond the obvious sorrow and loss, it was an ironic death for a young pilot who had survived bombing raids during the Great War, in the accident-prone DH4, only to die on a peacetime assignment, in the name of surveying.

A few weeks later, a flyer whom the records recall only as Pilot Grady and a co-pilot known only as Sergeant Le Seur completed the assignment in a Bristol Fighter, a more stable biplane, also from Base Borden. The last batch of photos enabled Nelles to create what Owen Thornton, a London writer, has called "the first complete aerial photograph mosaic of any Canadian city."

Because of Thornton's interest in the sad fate of Hubert Holland, the great aerial photo will now forever be associated with both Holland and Owen, who displayed the kind of courage without which aerial surveying in Canada could not have advanced as rapidly and functionally as it did.

PHOTOGRAPHY FROM THE AIR could certainly not have advanced fast enough for those conducting the legendary survey of the Ontario-Manitoba border during the early 1920s. The vast and dramatic project began under conditions of the sort that had driven many an honourable survey crew to drunkenness and despondency, if not mutiny. Paradoxically, by the time the survey was completed many years later, it had become a kind of object lesson in progress, amalgamating five-thousand-year-old transportation, in the form of horses and dog teams, with the latest in photogrammetry and winged transport.

CHARLIE FAIRHALL • YOUNG CHARLIE LOSES A CLIENT

In 1955, Charlie Fairhall, 26 years old and fresh from articling in Hamilton, went to work for the surveying firm of Fitzmaurice and Boyer, in Bracebridge, in the district of Muskoka. Just before Christmas that year, the firm was contracted to sever a waterfront property on a nearby lake into several cottage lots.

On December 21st Charlie and his assistant Len were sent to pick up the elderly owner of the property and take him to the lot so that he could advise them on the work.

The day, as Charlie recalls, dawned clear and very cold. The client was dressed cozily and carried a packsack and a stout walking stick.

"When the three of us arrived at the place where we had to leave the car," wrote Charlie in February of 2015, "I expressed a concern that, given the fresh fall of snow the night before, the long walk up the lake might be too strenuous for the old gentleman (for discretion, I'll call him "Mr. Brown"). But he assured us that he was up to the task and opined prophetically that if anything happened to him, his family would have a merry Christmas without him."

At the lake's edge, the old man indicated a quite-distant point of land, beyond which lay his property. He told the younger men to go ahead; he would follow at his own pace.

"And so, we proceeded, pulling a toboggan on which we'd piled our axes, transit, steel tape, iron bars, and so on."

Having rounded the distant point, Charlie and Len waited for Mr. Brown to catch up. When several minutes had passed and he hadn't come into view, Charlie ran back to look for him. "To my astonishment, as I came around the point, I could see in the distance the dark figure of our client, lying on the snow. 'He's down!' I cried out to Len and immediately ran to assist him. But it was too late. Mr. Brown's stout old heart had stopped and he was beyond help.

"When Len arrived, he was distraught, could hardly pull himself together. I told him firmly to return to the car, drive back to Bracebridge and fetch the OPP while I stayed with Mr. Brown.

"For some reason as I stood by the stiffening body, stamping my feet and clapping my hands, trying desperately to keep warm, lines from Robert Service's poem, 'The Cremation of Sam McGee,' kept popping into my head."

After what he describes as 'an eternity,' Charlie, now nearly frozen, spied Len and an OPP constable making their way up the lake. "They'd brought the coroner with them in the car," he said, "but the chap had refused to accompany them onto the ice lest he too expire making the effort. 'Hell,' he told them, 'you can tell if the man is dead!'

"I'll never forget the reaction of the constable when he arrived on the scene. Without even a glance at poor old Mr. Brown, who was now pretty much an ice cube, the cop walked up to the transit, fastened his eye to the telescope, and exclaimed, 'Gawdam, I always wanted to look through one of these!'

"In no time, we'd strapped Mr. Brown to the toboggan along with all the rest of our gear, and Len and I pulled the load back to the car. The constable excused himself from the chore on account of his 'bad back.'

"In town, we assigned Mr. Brown to the capable hands of the local undertaker, and in later years I often remarked that it was the first time a client of mine had suffered a heart attack before he got the bill!"

Charlie Fairhall received his surveyor's commission in 1954. Over the years, he practiced with a variety of companies in locales that included Bracebridge, Sudbury and North Bay. In 1970, he and partner Harland Moffatt started the Ottawa firm Fairhall and Moffatt, which later became Fairhall, Moffatt & Woodland.

Paul Wyman • Another Victim of the Homonym

During the summer of 1967, having completed high school in Kitchener at the age of 18, Paul Wyman settled to a year in the work force, during which he hoped to raise money for eventual entry into university. Driven by his parents' early deaths from leukemia and cancer, he dreamt of becoming a micro-biologist and doing medical research.

His acceptance into the University of Waterloo the following spring brought with it high anticipation for intellectual debate, expansive reading, and contact with professors who cared deeply about their work.

By November of his first term, however, he was not only severely disillusioned with the realities of academic life – huge classes, remote professors, intellectual apathy – but was also broke and hungry, his student loan having failed to materialize.

Thus it was that he quit and made his way to the local unemployment office.

"As the job officer leafed through the postings," Paul wrote in March, 2015, "he stopped at one and asked if I knew anything about doing a survey. I would have said yes regardless, but in fact I actually *had* filled in for a friend doing door-to-door and telephone surveys during high school, so I could say yes legitimately."

The following day, Paul made his way to the offices of John Metz OLS Limited on Frederick Street in Kitchener. "Mr. Metz made the usual inquiries as to my age and education and then asked me to solve a number of math problems – mainly geometry and trigonometry. Fortunately I have some facility in mathematics and managed to satisfy him as to my capabilities.

"As the interview progressed, however, I got increasingly uncomfortable. Clearly, the job was not about knocking on doors to ask the lady of the house about her favorite laundry detergent, or making phone calls to determine her favourite television shows. It was about *land surveys*, and I had only the vaguest notion of what that might entail. Still, I was desperate for a job, and in the end gratefully accepted the offer of $65 a week to meet this strange challenge."

Paul started work on a fiercely cold day in early December, 1968, as a field assistant – a "tail chainman," in the jargon of the trade. Appropriate to that distinction, he wore unlined cowboy boots, light jeans and a short jacket... and was of course frostbitten, nearly paralyzed, by 9:00 a.m.

"When we returned to the office late that first afternoon, I went in to see Mr. Metz, explained my predicament, and sheepishly asked if I might get a small advance on my salary to purchase warmer boots, coat and mittens. Mr. Metz lent me $20.00 – and so began my near-half-century career in land surveying.

"During the decades since, I have sometimes wondered if I would have been a good microbiologist. I do regret never going to university. But I have never regretted the misunderstanding that came from the two different meanings of the word 'survey'. Sometimes we get what we need by persistence and planning, sometimes we just need to ride the wave of random opportunity."

Paul Wyman, OLS #1344, received his commission as an Ontario Land Surveyor in 1973 and served as president of the AOLS in 1993.

The resilient and innovative land surveyor, John W. Pierce, OLS #392, was hired in 1921 to conduct the first Ontario-Manitoba boundary survey, a storied project that lumped along for more than 15 years and introduced aerial photography to Canadian surveying.

The survey was initiated because post-war mining had spread north along the unsurveyed boundary. Governments in Winnipeg and Toronto considered it crucial to know precisely which metals and minerals belonged to which province. The southern part of the boundary, running north from the U.S. border to Lake of the Woods, had been surveyed and established more than 30 years earlier by the federal government – but not without great controversy. Indeed, its establishment had caused violent battles between hastily appointed border cops from both provinces. Manitobans resolutely believed the boundary should have been set 250 miles to the east, near present-day Thunder Bay, giving Manitoba "room to breathe," as a Winnipeg reporter phrased it. Oft-drunken officers from both provinces are said to have lined themselves up along the new boundary, taking turns throwing one another into makeshift jails, on the presumption that each was trespassing on the other's sacred terrain. In one instance, near Kenora, a jailed officer from Manitoba is said to have burned an escape hole in the wall of the wooden hovel in which he was imprisoned, to have leapt from the hole, arrested the Ontario drunkard who had arrested *him* and, with help, to have pitched "the devil from Ontario" head-on through the burn hole into the charred jail.

At the initiation of the northbound survey, in 1921, the Manitoba government was still smarting over the perceived injustice of the border's location. On that account, it refused to pay its share of the cost of the survey.

Ontario and the Dominion agreed to go ahead anyway, and immediately hired one of Canada's most respected surveyors, John W. Pierce, who, while born in Quebec, possessed surveying licences in both Manitoba and Ontario. A prototypical Marlboro Man (mustachioed, lantern-jawed, nicotine-addicted), Pierce was perhaps the last surveyor to be men-

A survey crewman during the early 1900s required not just knowledge and a sense of adventure but a strong back and legs. In remote areas where a crew had no access to horses or dogs, a man could be expected to carry upwards of a hundred pounds, sometimes for many miles. Opposite page: An aerial mapping team in northern Ontario, preparing for takeoff in a Vickers Vedette, circa 1923.

tored, if briefly, by the fabled Alexander Niven. In 1909, Pierce furthered his wilderness experience by heading for the Porcupine, where he surveyed prospecting claims during the gold rush. But he left hastily for Pembroke in 1911 after a cataclysmic fire killed nearly 40 residents of South Porcupine, injured and disfigured dozens of others, and eliminated the tinderbox that was the town.

On the bitterly cold afternoon of March 13, 1921, Pierce led a party of 26 well-bundled technicians, line cutters and teamsters east from Winnipeg to a mustering point north of West Hawk Lake, from where he and his men planned to proceed due north, clearing and marking 180 miles of boundary.

It is the stuff of legend, not to mention of potential revolt, that the party took with them more than ten tons of supplies – nearly 800 pounds per man – the lot of it drawn over snowy bush roads by Clydesdales and Percherons, then by sled dogs as the northbound roads dwindled to trails.

The actual boundary-making, which began on June 6, was a more epic grind yet, necessitating the felling of thousands of spruce and jack pines, the traversing of dozens of unanticipated lakes and bogs,

and the daily packing and transport of tents, supplies and survey equipment. Plus of course there was the daily surveying itself, hour after hour, and the painstaking construction of stone and concrete monuments.

THE WEIGHTILY ENCUMBERED procession struggled north until mid-October, when Pierce decided that it was time to return to civilization, even though he and his crew had run and marked just 70 miles of their 180-mile assignment.

During the winter, it occurred to Pierce that if he could obtain aerial photos of the landscape ahead, he could better plan for traversing the lakes and swamps, and could, he imagined, pick up his pace. One of his difficulties all along had been an inability to guess how far to traverse around large lakes, whose dimensions he was unable to calculate from ground level. He had been painfully discouraged to realize that in working around the many bodies of water in the boundary's path he and his men had trudged some six miles laterally for every mile they had moved forward. "If it had been an east-west boundary," he once joked, "we would have been fine, would have run 400 miles of it in four months… Unfortunately, we were trying to go north."

In late 1921, desperate to speed things up, Pierce contacted the Air Board, Canada's first aviation control agency, and asked for help.

Curious about the boundary project and how aerial photos might be applied to survey work on the ground, the board obtained a Vickers Viking – basically, a flying rowboat – and sent it north from Victoria Harbour on Lake Winnipeg to take what photos its crew could get of the welter of lakes and marshes that lay along the line.

As Pierce had anticipated, those early aerial images were a transformative ben-

Mapping from the air. What the early aerial photographer saw (including another aerial photography crew) as he roared along in the nose of a Vickers Viking or Vedette, 5,000 feet above the lakes and forests of the north.

efit to him in planning the next phase of his task. Come summer, he also made good use of a simple traverse that had been run during the winter up the entire east side of the waters and wetlands ahead, meaning that he did not have to traverse each feature individually.

In early July of that second year, the Air Board's Viking flew into the survey camp with mail and supplies, allowing Pierce to get up into the air and discover for the first time what the territory looked like from above. While the aerial photos he was using were by no means high-quality, they were at least functional, and in a country as large and open as Canada they proved to be the beginnings of a vastly productive allegiance between aircraft, photography and the art of surveying.

Pierce and his men finished the 180-mile run to the Twelfth Base Line early in the autumn of that second season. However, Pierce did not proceed further with the survey until 1930, when aerial photos lightened his work to the extent that he was able to push the line another 90 miles toward Hudson Bay in just 17 days.

By that time, aerial photos were widely available to surveyors, and a kind of functional tango had begun between the Department of Lands and Forests (which continued to advance surveying in New Ontario) and the RCAF, which was busy doing aerial work on the Topographical Survey of Canada. For its part, Lands and Forests decided to "open up" a dozen or more wilderness base lines and meridians, including Niven's Meridian, so that the lines would be easier for pilots and photographers to locate from the air.

WHILE THE NOTION OF "OPENING UP" a base line so that it was visible from the air might sound simple enough, what it meant in reality was a preposterous expenditure of human time and energy. To create a six-foot-wide opening at treetop or "skyline" level required a swath thirty feet wide at ground level. Cutting even a mile of such swath could require the felling of several hundred trees, often old-growth spruce as big around as 40-gallon drums – this in the days before chainsaws. In a wilderness where a hun-

One of trustiest appurtenances of the early Canadian surveyor was the 66-foot Gunter's chain, invented in England in 1620. A "chain" became a unit of measurement and to this day represents, for example, the width of many Ontario roads and highways, including road allowance.

To clear line through the forest in the days before chainsaws required the chopping and felling, and sometimes the removal, of a hundred or more mature spruce or jackpines per mile, some of them up to two feet in diameter – this over hundreds of miles of terrain.

dred miles of line was the merest fraction of the total grid, battalions of men were kept busy chopping ten hours a day, six days a week, for months on end. Years of a life could be consumed in season after season of axe-swinging and buck-sawing, in rising daily at dawn, enduring black flies in summer and the savage cold of the Ontario north in winter – only to be forced to start again as saplings on the cleared ground embarked upwards, quickly filling the swath.

The requirement to keep wilderness crews nourished was in itself an extraordinary challenge (although much easier once airplanes could bring food to remote camps). During the Great Survey of 1900, team leader James Robertson noted that one of his crewmen "ate at dawn the fillets from a pickerel of 18 inches fried in lard, and then a quart of oatmeal with lard, and then half a dozen slices of camp bread with lard, as well as a half-quart of beans with lard." All of this he washed down with "numberless cups of tea" – presumably minus the lard. Robertson noted that the same crewman consumed an identical meal of high-calorie staples at sunset, "with renewed vigour and appetite."

It hardly needs saying that, except for the occasional female camp cook (often an Aboriginal woman), surveying in Ontario during the early 1920s was the exclusive domain of men. Even the idea of a woman becoming a surveyor in the province was at the time inconceivable, although several female surveyors held licences in the U.S. The AOLS membership did include a prominent surveyor with the somewhat ambiguous name, Shirley King. But nobody close to the profession was even remotely confused over Shirley's gender. By the simple syllogism "Shirley King is an Ontario surveyor; all Ontario surveyors are men," Shirley was a man.

Coincidentally, the man who completed and retraced the Ontario/Quebec boundary with Shirley King, beginning in 1930, bore the given name Marie. By the same incontestable syllogism, J. Marie Roy,

From 1921 to 1946, The Star Weekly, *Canada's biggest news magazine, featured "Birdseye Centre," a weekly cartoon by Jimmy Frise of Port Perry. In the autumn of 1934, an installment of the cartoon entitled "The Surveyors – ??" appeared and, with Frise's permission, was featured on the menu of the gala dinner at the Association's AGM in February, 1935. Frise, who was severely wounded at the Battle of Vimy Ridge, was said to be familiar with the wartime work of Canadian surveyors overseas. He is seen at left working in his studio.*

American women came to surveying earlier than their Canadian counterparts. The all-female crew seen below was photographed in Idaho in 1916.
At right: Ontario surveyor, Shirley King, and Quebec surveyor, Jean Marie Roy, teamed up to retrace the Ontario-Quebec boundary in 1930. They are seen here in their sub-Arctic business suits, shaking hands over the boundary's northernmost monument at James Bay.

QLS, was a man – and everybody knew it when Shirley and Marie went off into the woods to retrace the old boundary.

Perhaps moved by the Supreme Court decision in 1919 to give all Canadian women the vote, the AOLS made an attempt in 1922 to make women more welcome at its annual general meeting in Toronto. One can only imagine the deep tremor of excitement that must have gripped members' wives when they were invited to listen to a lecture on the topic, "Design Specifications for the Transportation Systems of Brazil and Venezuela." For reasons not entirely elusive to the imagination, the following year's lecture was cancelled and the women released to their own ideas of an afternoon's enjoyment.

By 1924, a "Ladies Committee" had been formed, leading to several years of noteworthy upgrades in entertainment at the association's AGM: museum and art gallery visits, "motor bus" tours of Toronto, theatrical productions, and, perhaps inevitably, shopping expeditions to Eaton's and Simpson's and Birks. In 1929, the women had tea at the newly opened Royal York Hotel – and the following year a special "women's dinner," including a tour of the hotel, with its 66-foot telephone switchboard (requiring 35 operators); its cavernous kitchen (still the largest scullery in Canada, with space for the fuselages of seven jumbo jets); its inside golf course (gone now); and its concert hall, where a pipe organ made by the famous Quebec firm, Casavant Frères, was then the largest in Canada.

The 1930s saw little change in the status of women in surveying or in any other Canadian profession. Indeed, nearly four decades would pass before the male citadel that was surveying in Ontario was penetrated by Canada's first female surveyor. For now, it is worth noting that during the winter of 1932, that very surveyor, Lorraine Gladstone, was born in Sudbury, where her father was a nickel smelter. Thirty-seven years later, having articled in Leamington to her husband Bill Setterington, Lorraine would gain her Ontario surveyor's licence, and from 1976 to 1992 would serve as executive director of the AOLS.

However, apart from the birth of its eventual executive director, the 1930s, with their desperate economics, held little promise for surveyors, or for anyone else in most parts of the country or continent.

Always the Black Fly No Matter Where You Go

Two of the more recognizable inhabitants of the recorded history of northern Ontario are the durable wilderness land surveyor and the unendurable black fly. Several early Ontario paintings or drawings feature a tormented-looking surveyor standing in a forest or swamp, surrounded by a swirling rout of tiny black specks (specks known collectively in some parts of the province as the Timiskaming Air Force).

However, it is in a piece of music, a beloved Ontario folk ballad, "The Black Fly Song," that the little winged bloodsucker, despised by surveyors, has found its greatest and most spirited fame.

Ontario surveyor Al Worobec tells the story of the survey crew that slogged into the wilderness north of Cochrane during the summer of 1949 to take the measure of the Abitibi River and its watershed, on which The Hydro Electric Power Commission of Ontario planned to build a pair of dams.

Al's father, Tony Worobec, from Barrie, was an instrument man on that crew, under the direction of crew chief Toby Colpits.

It was at some point that summer, amidst the spruce forests and bogs, that geography and music met entomology in the form of the little black crazy-maker, *Simulium yahense*.

The musician in the group was a young song writer and painter named Wade Hemsworth, who eventually converted the summer's experience into the aforementioned folk song.

In 1956, for reasons obscured by time, the Hydro Electric Power Commission hired the artist H.H. Leering to do a painting of three members of the crew that had worked on the Abitibi River

survey. Titled *Survey Crew*, the work (facing page) depicts, from left to right, chainman Phil Thompson, instrument man Tony Worobec, and chainman Tom Letang, all of whom worked at times under Toby Colpits, the infamous "Black Toby" of the song.

Eventually, the Hydro Commission presented Tony Worobec with the painting, which now hangs in the offices of Eplett Worobec Raikes Surveying Ltd. in Barrie.

The song, meanwhile, has been sung for decades at campfires across North America. More famously, it was part of the repertoire of the American folk singer, Pete Seeger, and also of the renowned Canadian act, The Travellers, and of Kate and Anna McGarrigle.

In 1991, it was featured in a National Film Board animated production called *Blackfly*, which was nominated for an Academy Award and can be seen on YouTube.

Right: *Wade Hemsworth serenades (from left) Kate McGarrigle, Adrienne Clarkson, and Anna McGarrigle.* CBC archives.

Left: *This ink drawing from 1786, entitled* A Surveyor and His Struggles *and featuring a torment of black flies, was used as the frontispiece for the published report of the first meeting of the Association of Provincial Land Surveyors in 1886.*

H.H. Leering's luminescent painting, Survey Crew, *(facing page) was created in a style similar to that of the artists of the early Soviet Union: a highly realistic, almost photographic, bravura that honoured working people and tended to present them as motivated, forward-thinking and gloriously proud of their contribution to the utilitarian cause. Like the three surveyors in Leering's work, Soviet subjects were often depicted as "on the march" in the interest of the working and professional classes.*

The Black Fly Song (excerpt)

by Wade Hemsworth

'Twas early in the spring when I decide to go
For to work up in the woods in north On-tar-i-o.
The unemployment office said they'd send me through
To the Little Abitibi with the survey crew.

CHORUS: *And the black flies, the little black flies*
Always the black fly, no matter where you go.
I'll die with the black fly a-picking my bones
In north On-tar-i-o-i-o, in north On-tar-i-o.

Now the man, Black Toby was the captain of the crew
And he said, "I'm gonna tell you boys what we're gonna do.
They want to build a power dam and we must find a way
For to make the Little Ab flow around the other way."

So we survey to the east and we survey to the west
And we couldn't make our minds up how to do it best.
Little Ab, Little Ab, what shall I do?
For I'm all but goin' crazy on the survey crew.

It was black fly, black fly everywhere,
A-crawlin' in your whiskers, a-crawlin' in your hair,
A-swimmin' in the soup, and a'swimmin in the tea.
Oh the devil take the black fly and let me be.

Queen Elizabeth Way near Jordan Harbour, *oil on canvas, by Moma Markovich, 1966. Government of Ontario Art Collection, Archives of Ontario 101238.*

CHAPTER FIVE

MAKE WORK & MAKE IT QUICK

THE DEPRESSION AND THE AOLS

TO GET EVEN AN INKLING of the effect of the Great Depression on the surveyors of Ontario, one might consider that in 1931, all 91 surveyors employed by the Department of the Interior in Ottawa were given an abrupt notice of dismissal. At the Ontario Department of Lands and Forests, the staff of surveyors dropped from eighteen to four. The Surveys Branch of the provincial Department of Highways was the only government agency that continued to employ a decent, if fluctuating, contingent of surveyors (anywhere from eight to fifteen). At private survey and engineering firms, rosters of four or five surveyors were pared to one or two. Or none. Some firms simply quit, or went dormant, rather than waste money on a phone that no longer rang and on office space where no business was being done.

Unlike life in any other decade since the incorporation of the AOLS, there was, by late 1931, no axiom for survival – and no wreckage to cling to. Nobody was creating or expanding businesses, farms or factories. Urban growth had vanished, as had any planned expansion of shipping or transportation in the province. Whatever hydro projects were being pushed forward against the flow of economics had already passed the stage at which significant efforts in surveying were required. To run a trickle of money into the pockets of Ontarians, the provincial government of Premier George Henry funded several high-profile road projects. One was the construction of the Queen Elizabeth Way, the country's first four-lane highway, eventually opened in 1939 by King George VI and Queen Elizabeth, while another extended Highway 11, then the longest highway in Ontario, along its costly wintry path into the northwest.

The province's make-work projects included the famed Botanical Gardens, in Hamilton, which were surveyed, contoured and planted in the name of "provincial beautification." In Toronto, the Ministry of Culture began a reconstruction of historic Fort York, which

NORMAN WILSON • UNDERSTATED VIRTUOSO

"If at first the idea is not absurd," said Albert Einstein, "then there is no hope for it."

In February, 1932, there was certainly no hope for surveyor Norman Wilson's "absurd" recommendation to the Toronto Transit Commission that they consider building a subway beneath Yonge Street in Toronto at an anticipated cost of $12.5 million.

Wilson, who at the time owned a survey-engineering firm with Alexander Bunnell, possessed a genius for rail design and urban planning and would eventually design transit systems as far away as Mexico City, Rio de Janeiro and Sao Paulo, Brazil. In Canada, he would create transit plans for 18 cities, coast to coast. However, in his hometown of Toronto, during the harshest years of the Depression, his subway plan was not only ignored by the TTC but was ridiculed in the press, despite the prospect that it would create jobs for hundreds of destitute Torontonians. The Toronto Star wrote:

Wilson and Bunnell
Would build a fine tunnel,
But isn't it funny
That Toronto lacks money
To pour down a funnel.

Wilson had the pleasure of quoting the lines back to their author during the celebration to open the subway in 1954 – a subway that used Wilson's 1932 design with barely a deviation but that was built at many times the cost of his original estimate.

Born in Toronto in 1883, nine years before the founding of the AOLS, Wilson was admitted to the Association in 1910, served as its president in 1944, and during a five-decade career accomplished

as much as perhaps any surveyor in the annals of the AOLS.

As chief surveyor for the Toronto Harbour Commission in 1913, he carried out the first detailed survey of the Toronto waterfront – an initial step in the development not just of the modern port of Toronto but of recreational grounds for the growing city, and of the drainage of the vast marshes at Ashbridge's Bay where the Don River meets Lake Ontario.

During World War II, Wilson planned and supervised all Canadian transit systems related to the war effort. During the same period, as AOLS president, he campaigned against "the stupidity of Ontario's checker-board township survey system." During his presidential address to the Association in 1944, he asked, "How many hundred-acre farms have been cleared when actually only a fraction of each was agricultural land?"

Where mining was concerned, Wilson advocated claims covering mineralized areas only, not arbitrary 40-acre blocks. And he deplored how historic lot lines dictated the Toronto street system, which he pointed out ran "diagonal to the topography," resulting in "enormous additional costs in city development."

During his year as AOLS president, Wilson also completed the "Wilson Report," an intricately researched study of the human and geographic effects that would result from the building of the planned St. Lawrence Seaway, one of history's grandest feats of engineering. His report mapped the projected spread of water from the damming of the St. Lawrence River at Cornwall – an inundation that would make navigation on the Seaway possible at the same time as it established an immense international hydro project. The report suggested relocation plans for flooded highways, villages and cemeteries, and recommended sites for deep-water docking facilities – all of which led in 1951 to the creation of the St. Lawrence Seaway Authority which endures today.

During the early 1950s, Toronto named Wilson Avenue for this understated virtuoso, as well as the area known as Wilson Heights, and later Wilson Station, a multi-use transit facility at Wilson Avenue and Allen Road. Wilson died during Canada's Centennial year and was remembered during the Association's special celebrations in 1967 as "a man whose unassuming ways and quiet voice gave no indication of his great intelligence, integrity and force of character." It was said of him at his funeral that, as a surveyor, he "observed closely, calculated impeccably, and applied his imagination."

Right: Numerous Toronto surveyors participated in Norman Wilson's massive project, begun in 1913, to build docks and recreation areas along Toronto's eastern harbour.

Below: The Edward L. Ryerson, *launched in Wisconsin in 1959, has long been considered the most aesthetically pleasing ship on the Great Lakes. During the early 1960s, the 730-foot American-owned ore carrier was also the largest vessel on the lakes. She is seen here leaving Lock #2 of the Welland Canal, a crucial section of the St. Lawrence Seaway.*

The reconstruction of Fort York in Toronto and the creation of the Royal Botanical Gardens in Hamilton, seen below, were Depression-era make-work projects, each of which provided a few paydays for underemployed surveyors.

meant occasional employment for a surveyor or two and a crew of chainmen and technicians.

Surveyors in the north got a whiff of opportunity when the federal government passed the Relief Land Settlement Act, through which the province, taking a final run at establishing farms in the frost belt, selected some 600 luckless souls from depressed urban areas and "resettled" them along the railway tracks in the districts of Thunder Bay, Hearst and Kapuskasing. Two hundred or more 100-acre farms (most bearing a mysterious likeness to 100-acre forests) were surveyed in the process.

However, as always, any attempt to farm the unfarmable bore little in the way of a harvest. By 1939, more than half of these agricultural enterprises, many of which had never been properly cleared, were derelict and abandoned. Of those still inhabited at the end of the Depression, only a fraction were being ploughed and seeded.

Like so many determined citizens without work, a number of Ontario surveyors and their unemployed crewmen are said to have boarded west-bound trains and followed the tracks, picking up a few dollars here or there on rail or road surveys.

It is also possible that a handful of young surveyors or crew – or perhaps articling students desperate for something to do in the collapsed economy – found their way into the province's relief camps. The latter were frightful places gouged out of the northern wilderness under the sponsorship of the National Department of Defence. Surveyors or crewmen who were not residents of the camps, and perhaps some who were, may actually have found the occasional day of paid (or unpaid) surveying in, say, the road-building or hydro-line cutting, or the construction

Left: An all-purpose dorm in one of the hated Ontario relief camps during the early years of the Depression. Below: Camp inhabitants – some would call them "inmates" – digging a road bed in their bare feet for two cents an hour.

of public buildings, to which the camps were dedicated.

What is perhaps most intriguing about the relief camps in the context of these paragraphs is that they were intricately connected to another chapter in the story of surveying in Ontario. For it was Major General Andrew McNaughton – he of heroic reputation among the surveyors of World War I, and now Canada's military chief of staff – who declared that the country's "idle" young men should be put to work before they "took to communism" and fomented a revolution against the government of R.B. Bennett. "In their ragged platoons," said McNaughton, "they are the prospective members of what Marx called the 'industrial reserve army, the storm troopers of the revolution.'"

Once established in a camp, the young men – as many as 40,000 of them in Ontario – were fed and housed, at more or less prison standards, and were paid two cents an hour for ten hours of work a day. They were forbidden from organizing politically and, against their bitter protests, were denied a vote in the federal election of 1935.

It is hardly surprising that in the end the camps were ineffective. Rather than diffusing political tension, as McNaughton had predicted, they inflamed it. "In those bunkhouses," wrote labour historian Ronald Liversedge, "there were more men reading Marx, Lenin and Stalin than there were reading girlie magazines," which were described by one wag as "the official literature of the camps."

In all, such places were an unfortunate but telling reflection of the man who had plotted so painstakingly during the war to preserve the democratic freedoms he now seemed intent on denying.

Meanwhile, the Depression was as hard on the AOLS as it was on individual surveyors. During the mid- and late 1930s, virtually no new surveyors came into the profession. Many members either couldn't, or refused to, pay their dues. By 1939, the membership had shrivelled to a twentieth-century low of about 200.

It was during these dark days that the AOLS lost a pair of its founding heroes: Willis Chipman, who had done more than anyone to create the Association, and Elihu Stewart, who had goaded the province to get moving on the Great Survey of 1900. Chipman died in 1929, Stewart in 1935. A buccaneer to the end, Stewart conducted his final survey in the forests of Nottawasaga Township, in 1931, at the age of 87.

As bad as things were, the Association did rouse itself in 1936 to celebrate its 50th anniversary (a date measured not

The Royal York Hotel in Toronto hosted the 50th anniversary of the AOLS in 1936. Just 96 surveyors attended the muted celebration, whose gala dinner was brightened by a one-hundred dollar contribution from the City of Toronto. Below: General Andrew McNaughton, second from left, ran Canada's Depression relief camps. But with the advent of war, he was back in Europe where he had led the country's surveying and engineering effort in World War I and would now have broader responsibilities. He is seen here with Winston Churchill, Charles De Gaulle and Wladyslaw Sikorski, the exiled prime minister of Poland.

from incorporation in 1892 but from the unincorporated founding in 1886). Even with membership in decline, and largely in the absence of men under forty, 96 trail-weary surveyors gathered at the Royal York Hotel in Toronto, in late February, to talk and dine and dance – and, assuredly, to complain: about everything that had conspired to undermine their lives and bank accounts and profession.

The City of Toronto brightened the picture by contributing one hundred dollars, considered then to be a very generous sum, to the golden-anniversary dinner and entertainment. It was an event at which the tables, with their fresh flowers (and presumably not-so-fresh faces), were reported in the press the next day to have been "strikingly colourful and gay."

A church service on Sunday morning is said to have concluded with prayers for delivery from what was now nearly seven years of tattered ambitions and bottomless bottom lines.

LITTLE DID THOSE PRESENT KNOW that the help for which they prayed would soon be on its way, not from above but from below, and in a form more twisted and catastrophic than anything they could have imagined. What no Depression-era government could achieve with make-work projects or relief camps, the declaration of World War II achieved in short order in 1939: full employment.

Great Britain had no sooner declared war on Germany in September of that year than Canada declared its own battle intentions, and thousands of young Canadians signed up, took basic training, and boarded troop ships bound for England and France. The recruits included 24 Ontario land surveyors, several dozen survey technicians and crewmen, and a dozen or so students whose articling ambitions had been stretched, if not entirely eclipsed, by the Depression. If the number of recruits seems low (particularly compared to the flood of surveyors who went to war in 1914), it is because so few men of enlistment age had come into the profession during the previous decade.

It is worth noting that dozens of young Ontario men who went overseas with no surveying credentials had their interests piqued by engineering, mapping and flight responsibilities during the war, and became surveyors when they returned home. Among them were the

This graphic pitch to potential enlistees appealed to their sense of chivalry, daring and speed. The solid Art Deco style suggested Canadian strength, perhaps even invulnerability.

Guelph surveyor, Bill Carr, who served as a navigator with the Allied Flight Command; and the Stratford surveyor, Fred Pearce, famous for spending three years in a German prison camp and escaping on a forced march near Hanover a few days after Hitler announced that all prisoners-of-war were to be executed; and the London surveyor, Jack Gray, who served as a sapper with the Royal Corps of Engineers, preparing roads and bridges critical to the final liberation of France and Holland; and "Red" Petzold, who piloted Lancaster bombers with the Allied Flight Command throughout the war; and Harvey Hall, who began the war with the Royal Corps of Engineers and finished as Sergeant Major of the Second Canadian Field Survey Company.

Neil Simpson, another eventual surveyor, aimed bombs aboard a Stirling MK4, on night missions over Germany. "As a kid, during the Depression," Simpson said in an interview in 2015, "I'd worked with surveyors building the Rainbow Bridge at Niagara Falls. I always knew that if I survived the war I'd go back to surveying." Which is what Neil did, gaining a reputation as a kind of boreal cowboy, operating out of North Bay, piloting his Cessna 180 from one job to another across the north for some 40 years.

The legendary Ontario surveyor, Colonel Sam Archibald, was active in both world wars, returning from World War I to take up surveying first in Sault Ste. Marie during the early 1920s, then in the London area, in 1925. While in the Soo, Archibald met and befriended Dane Wandabense, who at that point (and for many years to come) was the only Aboriginal surveyor in the history of the AOLS. In part because of Archibald's influence, Dane would go on to serve as a sergeant with the Canadian

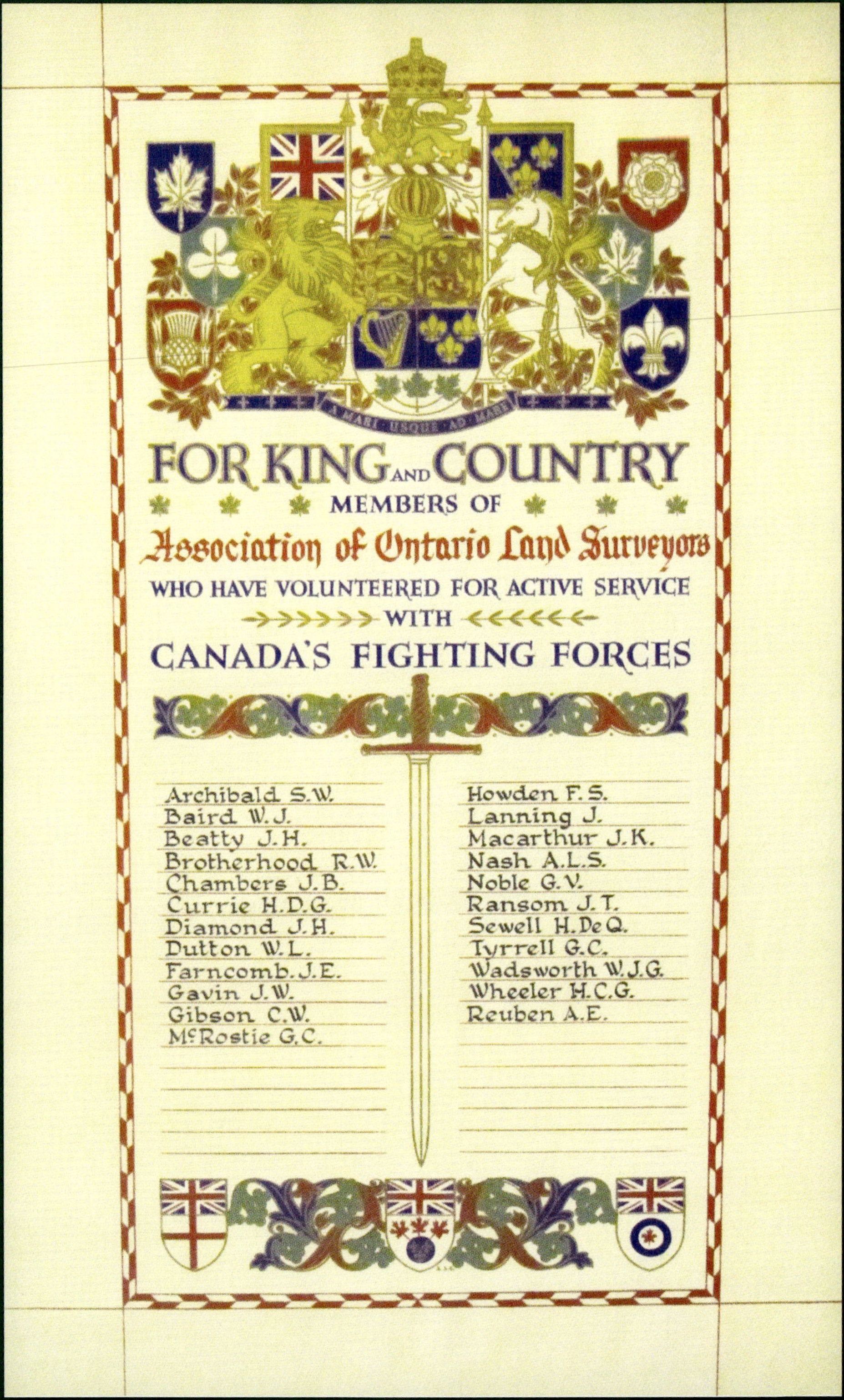

A MARI USQUE AD MARE

FOR KING AND COUNTRY

MEMBERS OF

Association of Ontario Land Surveyors

WHO HAVE VOLUNTEERED FOR ACTIVE SERVICE

WITH

CANADA'S FIGHTING FORCES

Archibald S.W.	Howden F.S.
Baird W.J.	Lanning J.
Beatty J.H.	Macarthur J.K.
Brotherhood R.W.	Nash A.L.S.
Chambers J.B.	Noble G.V.
Currie H.D.G.	Ransom J.T.
Diamond J.H.	Sewell H.DeQ.
Dutton W.L.	Tyrrell G.C.
Farncomb J.E.	Wadsworth W.J.G.
Gavin J.W.	Wheeler H.C.G.
Gibson C.W.	Reuben A.E.
McRostie G.C.	

This hand-lettered honour roll names the 23 Ontario Land Surveyors who volunteered to serve in Canada's armed forces during World War II. Many more young men who fought in the war (often doing surveying or mapping work) became Ontario Land Surveyors upon returning home after the European armistice in 1945. This framed document hangs in the AOLS offices on McNicoll Avenue in Scarborough.

Sam Archibald

W.J. Baird

H.D. Currie

J.H. Diamond

W.L. Dutton

C.W. Gibson

Jack Gray

George Noble

Fred Pearce

John Peters

Lloyd Peters

"Red" Petzold

Opposite page: A Lancaster bomber, guided by exacting instructions from England, on a mission over Europe during World War II.

Left: Canadian enlisted women engaged in map-making from aerial photographs in 1944.

Right: RAF pilots preparing for a bombing run.

artillery corps during World War II before returning home to article as a surveyor under the famed Harold Howden of the province's Department of Highways.

AMONG THOSE WHO WERE ALREADY surveyors when they put on a uniform, several joined artillery units overseas, while others took up aerial photography and map-making, or learned the intricate skills required to assess topographical maps and terrain, and to plot effective attack routes and bombing raids.

Not surprisingly, map-making from aerial photos became something of a specialty for the Ontario surveyors, many of whom had links to the Geodetic Survey of Canada, or were otherwise familiar with airborne photography and the conversion of the photos into maps. As always, the imperative of the surveyor at war (whether under Genghis Khan, Field Marshall Montgomery or Major Andrew McNaughton) was that "an army fights and wins as much on good maps as on good generals, weapons and strategy."

As the war ebbed and flowed, Canadian map-making units became increasingly vital to the effort, moving as they did along the allied lines, in two types of specialized trucks or trailers. One of these housed surveyors and cartographers, whose job it was to convert aerial photos into functional topographical maps, while the other housed large-format map-printing presses on which the maps could be printed on demand. The two were never far apart, and there was no rest for either, day or night. Photos taken on Monday, for example, often had to be converted to maps for distribution to the field units on Tuesday. In addition, maps were constantly being redone, upgraded, or expanded. Trucks carrying as many as a thousand different maps, often in multiples of a thousand copies or more, worked the lines, supplying the latest in cartography to those who were firing the artillery cannons, planning the next assault or air strike, or dropping the bombs.

BACK HOME, ONE OF THE GREAT clandestine stories of the war – a tale of persistent surveyors, secret knowledge, and a vast and brilliant logistical effort – unfolded deep in the forests of northwestern Ontario. Iron, it should be said, was crucial to the allied war effort, in that everything from trucks and tanks to ships' hulls, artillery cannons, and ammunition was made of steel from smelted iron. The problem was that the British and Americans, manufacturers of most of the machinery and armaments for the war, simply did not have enough of the vital ore to satisfy the proliferating need, either military or domestic. South American iron mines had long supplied iron to the two countries, but from 1940 onwards German U-boats torpedoed and sank more than 85 percent of the ore ships leaving seaports such as Belem and Valparaiso.

The undeveloped Steep Rock ore body was by no means the most accessible or obvious new source of iron for the war

Dane Wandabense • One-Man Band

In 1979, at the age of 80, Dane Wandabense, a land surveyor from Little Current, on Manitoulin Island, took on an articling student, Bob Halliday, a recent graduate of the Erindale Survey Sciences Program.

Wandabense, who died in 1995, was, at the time, the only Indigenous land surveyor in AOLS history.* "He told me his birth records had been burned but that he knew he had been born in about 1900 on the Sheguianda Reserve, south of Little Current," Halliday said in an interview in September, 2015. "His parents died when he was five or six – nobody knew how; and he was sent to a Jesuit orphanage in Sault Ste. Marie."

At twelve, having reached what the Jesuits considered the age of independence, Dane, an intelligent and durable child, already nearly six feet tall, landed a job working for Thaddeus Patten, the renowned pioneer surveyor. Patten had plotted a host of wilderness townships along the upper shore of Lake Huron and had explored and surveyed numerous logging and mining tracts further north.

Thaddeus and his young assistant travelled to job sites by boat along Huron's north shore, often moving inland by train or canoe – or sometimes simply bushwacking their way through the spruce forest, using machetes and axes. "Dane's job was carrying and stamping the brass caps that they put on the monuments," said Halliday. "Each cap sat atop an 18-inch pipe with a foot plate that they'd dig into the ground or bury in a stone cairn."

When Patten died during the Depression, Dane took up prospecting, hoping to eke out a living. At the outbreak of World War II, in 1939, he joined the Canadian artillery corps and was sent to Europe. There, like so many Canadians with surveying experience, he was put to work spotting enemy cannon fire and making target calculations for the Canadian artillery.

"He was injured late in the war," said Halliday, "and during his convalescence, I think in Hamilton, he met this miniscule, animated Scottish woman who everybody called Gretchen even though her name was Windabel, and she and Dane got married."

It was while working for the Ontario Department of Highways during the late 1940s that Dane began formal training as a surveyor and eventually articled to the celebrated director of the department, Harold Howden.

On a memorable winter night in 1948, while working on the Trans-Canada Highway, Dane travelled by train from the Sault Ste. Marie area to a construction site north of Lake Superior and disembarked at an unofficial stop in the wilderness, several miles from his destination. "He was probably snowshoeing," said Halliday, "and before he knew it a pack of wolves had fallen in behind him at close range – undoubtedly hungry in mid-winter. He was a very tough and savvy guy in the wilds and did make it to camp, but I suspect if he'd stumbled or shown any fear the wolves would have been on him and would have finished him off, probably leaving very few traces."

Dane was awarded his surveyor's commission in 1950, and went into business in the rather unlikely locale of Port Credit, under the name James and Wandabense. According to biographical notes written by the surveyor Kenneth McConnell in the late 1990s, there was never actually a "James" in the business. Wandabense did have the notes of D.D. James, a long-dead surveyor, and is presumed to have added a little extra oomph to the name of his business by implying that two surveyors were involved.

"By the time I met him to start my articling," said Halliday, "he'd moved his office to Little Current, then to Espanola. Originally, I thought he was Danish – that "Dane" was a nickname of some sort. When I visited his home in Little Current for the first time, the door opened and here was this imposing Aboriginal man, eighty years old, standing beside this fiery little Scottish woman with a bright shock of white hair – a woman clearly in charge of both the house and Dane."

Halliday articled for a year with Wandabense, but moved on when the senior surveyor decided to retire in 1980. "It was a strange year, but fascinating," said Halliday. "Dane had lost his hearing during the war, from being so close to the artillery cannons. So almost all of our communication in the office and the field was in written notes, which of course tended to cut out small talk. We were really quite

---

* *Métis surveyor Francis Cadeau was commissioned in 2002.*

Bob Halliday OLS #1556, was granted his commission as an Ontario Land Surveyor in 1984. During the years that followed, he worked with T.E. Rody and W.J. Keatley, and from 2004 to 2013 ran an independent office in Espanola, where he is now affiliated with Tulloch Geomatics Inc.

isolated from one another. But when we tried to talk we'd mess up; he couldn't hear well enough to understand my questions.

"Otherwise, at eighty, he was still a very strong, vital guy. He'd come out with me on jobs where there was line-cutting to do, and he'd just be whaling away with his ax, knocking over trees. I was twenty-two and I had a hard time keeping up with him."

Dane was notably non-political about his First Nations affiliation and, according to Halliday, never mentioned the fact that his predecessors in surveying had played a sometimes nefarious role in the acquisition of land from the First Nations and the herding of Native bands, including Dane's own band, the Sheguiandah, onto reserves during the late 1800s.

"Even in the light of that history, I don't think his being a surveyor struck him as any sort of irony," said Halliday. "You've got to remember, he also fought for Canada and the King in the war. He was a survivor, not an activist. It might have been different if he'd been born fifty years later."

Dane did maintain close connections with the Sheguiandah Band and often hired young men from the local reserve as survey crew. He died in a nursing home in Espanola in 1995 and is buried under a granite headstone, beside Gretchen, in the cemetery in Little Current.

Thadius Patten's survey plan at right was certainly no "paddle-by." The surveyor noted in the plan's lower corner: "I hereby certify that this plan is correct, and is prepared from an actual survey performed under my personal supervision, and that I was in my own proper person present on the ground during the progress of such survey. – Little Current, Ont. August 3, 1926."

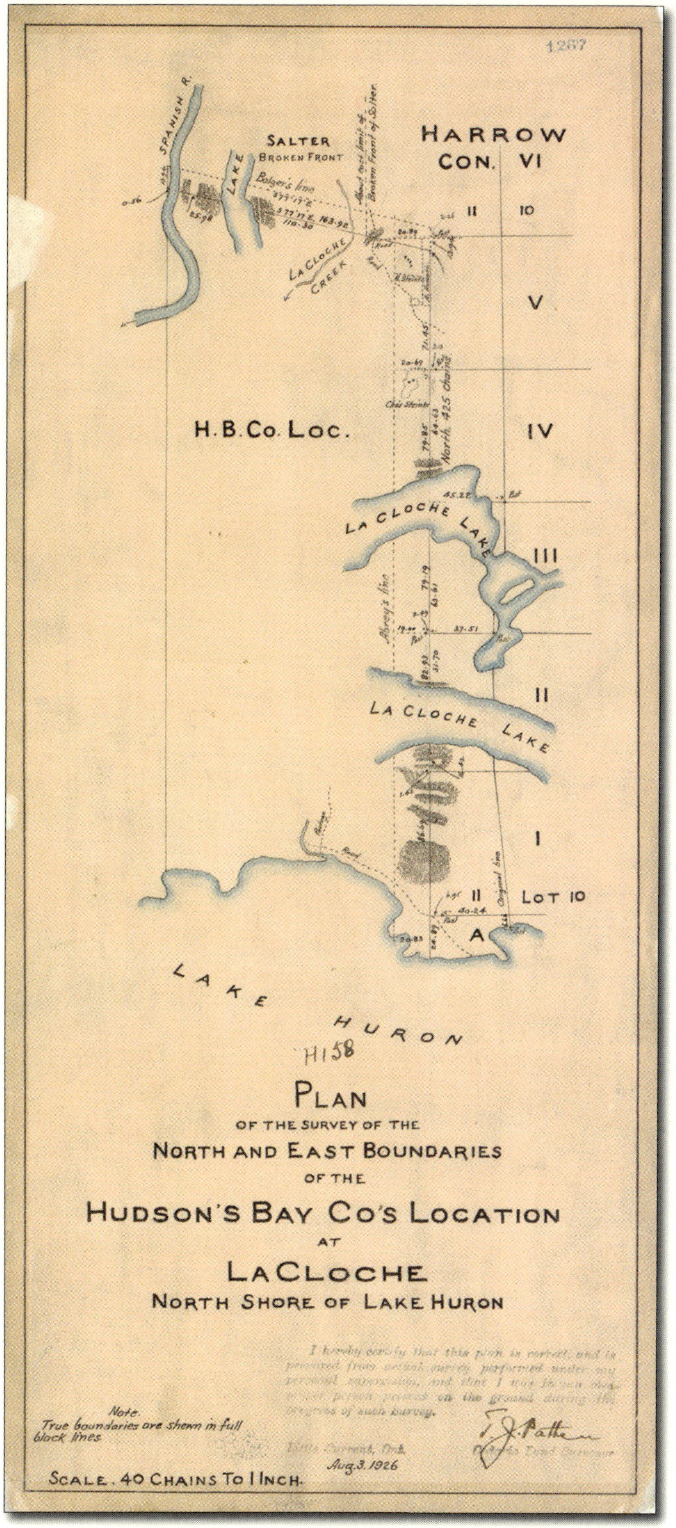

Left: Surveyor William Samuel assessing the shoreline of Wagita Bay on Steep Rock Lake in 1943. Opposite page: Finlayson Cut, 25 metres deep, showing one end of the 400-metre-long tunnel that would divert the Seine River around Steep Rock Lake.

Fort William's Roy Kirkup, OLS #480, surveyed the first mining claims at Steep Rock Lake in 1938 and, in 1943, with fellow surveyor Sydney Hancock and engineer, M.S. Fotheringham, drew up plans for the epic project that made the Steep Rock Iron Mine possible.

effort. Sitting about a hundred kilometres west of Port Arthur and Fort William, it had been discovered in 1897 during the Canadian Geological Survey. But it had attracted little attention and no development money from potential investors. For one thing, it lay entirely beneath Steep Rock Lake, which was about a hundred metres deep and was frozen much of the year. The ore body had never even been properly assessed until local engineer James Cross made a magnetic survey through the ice during the winter of 1930. Excited by his findings, he rounded up enough backing to establish a small exploration company and in 1938 drilled a number of test holes deep into the ore. "The results," said Cross at the time, "are far beyond anything any of us had anticipated – this is a huge and important body of ore."

During the same year, Roy Kirkup, a Fort William surveyor who at the time was president of the AOLS, was contracted to survey the first mining claims for what a year later would become the Steep Rock Iron Mine. His assistant, another regional surveyor, Sydney Hancock, who had articled to Kirkup and who in 1972 would himself become AOLS president, immediately took on a significant share of the surveying responsibility. Still, investors were reluctant to commit to an ore body, even a very rich one, that lay buried beneath 800 million tons of water.

But the war changed everything – and quickly. Within days of its outbreak, there was a brisk new demand for iron, the value of which rose precipitously. Investment in Steep Rock followed – first from the Canadian-born billionaire, Cyrus Eaton, then from the U.S. government.

The challenge of course was that the ore was still buried beneath nearly a billion tons of (albeit fluid) overlay. In order to bring it to the surface, the Seine River, which flowed into the lake, would have to be diverted. Then the lake would have to be drained – a task beyond anything ever faced in the realm of Canadian surveying and engineering. To accomplish it, Kirkup, Hancock and the others decided that two 20-metre-deep channels would have to be cut through 600 metres of solid rock. A 400-metre-long tunnel four metres in diameter would have to be drilled through the same dense basalt and granite. Several million tons of gravel

would have to be excavated.

Could it be done?

Nearly three years of strategic and technological effort would be required to find out.

But find out they did. By mid-June, 1944, after an exertion to rival the great engineering feats of history, the channels had been cut, the tunnels dug, the gravel excavated. What's more, 120 billion gallons of water had been pumped out of Steep Rock Lake and the dredging of the lake bottom had begun. Sydney Hancock had by this time been appointed the project's chief engineer and surveyor.

Down they ploughed into the muck and silt, pushing it into mountainous piles alongside.

Finally in late July, crews sighted ore for the first time, and the mine, a vast open pit, was operational.

In the months leading up to that exciting but unpublicized occasion, an ore-loading dock had been built in Port Arthur, on Lake Superior, a hundred kilometres to the east. Given the blanket of security that shielded the mine's development and production, even residents of the area were kept somewhat in the dark about the output of ore and precisely where it was going, as increasing numbers of ships began moving in and out of the harbour. For those who knew what was happening, the cautionary wartime rhyme was the order of the hour: "Loose lips sink ships."

For the better part of a year, iron from Steep Rock was smelted and melted, cast or stamped, into domestic goods and into every steel weapon and conveyance of war.

It is said that the Germans never discovered exactly where the iron was coming from. They did realize, however, that Port Arthur was the site of shipyards that turned out a significant number of corvette patrol craft – the last of which may have contained Steep Rock iron that had been converted to steel in southern Ontario or the northern U.S.

No one has ever known precisely the effect the mine had on the uncertain evolution of the war. In some phases of that violent history, it undoubtedly made a difference.

What we do know is that on September 2, 1945, to everyone's relief, the war

Left: Billionaire Steep Rock investor, Cyrus Eaton (in cap), chatting with mine president and chief engineer, M.S. "Pop" Fotheringham.

Below: Surveyor, Sydney Hancock (third from right), and chief engineer, M.S. Fotheringham (second from left), with the senior staff of Steep Rock Iron Mine, circa 1945.

Opposite page: Steep Rock Iron Mine in operation during the 1950s.

was over. The allies had prevailed. A new chapter began – both in the life of the mine and the life of the province and country.

Years later, the Port Arthur *News Chronicle* would report that "an Ontario land surveyor, Sydney Hancock, along with engineer and mine president M.S. Fotheringham," had been "the operational intelligence and imagination behind this huge and vital project."

More cynical observers have noted that, as in all wars, many died while investors and manufacturers got rich on the sale of resources and armaments – including those originating at Steep Rock.

In 1946, Hancock was named superintendent of mining operations at the site, and eventually became assistant to the president of Steep Rock Iron Mines. During the late 1940s, Hancock took personal charge of surveying the streets and building lots for the town of Atikokan, which for more than thirty years would be the social and bedroom community for the thriving mine. Hancock eventually became the town's reeve.

Finally, in 1980, the ore ran out. The pumps that had kept the open pit dry since the war were turned off. Water flowed back into the crater, eventually filling it to the brim. The conversion of the site to recreational use during the years that followed was a project that consumed the attention and intelligence of a whole new generation of engineers and ecologists, as well as a handful of Ontario land surveyors.

Peter Moreton • All the World's a Stage

The Stratford Shakespearean Festival is one of Canada's best-known, best-attended and best-loved annual celebrations. Beyond the slate of each year's productions and the actors who bring them to life, the heart of the annual extravaganza is the 60-year-old Festival Theatre, a throwback to Elizabethan pageantry. There, over the years, some of the world's finest actors and actresses have displayed their craft: Alec Guinness, Maggie Smith, Peter Ustinov, Jessica Tandy, Christopher Plummer, Martha Henry, and many more.

"Back in the mid-1990s," said Peter Moreton in an interview during the summer of 2015, "the management of the festival decided to revamp the theatre, which I suspect hadn't been altered since it was built in the 1950s. They wanted to change the angle of the curtain wall, in order to eliminate a blind spot for people who were seated in the front corners. Our firm was hired for the project.

"What you have to know is that the stage and the bowl of seats are configured in a circle around the precise centre point of the theatre. So it was that point that we had to identify before we could proceed with any adjustment to the seating and stage walls.

"The contractor went right at it – ripped out the seats on the lower level, tore up the floor; there were Bobcats going in and out as we worked.

"One of the really interesting historical aspects of all this is that when they started the festival back in the early 1950s, the theatre was just a huge concrete pad, with banks of seats and an immense canvas tent over it all. A very fine surveyor named Fred Pearce, who was our predecessor in Stratford and owned the firm that we bought, had laid out the pad for the tent and configured the original stage and seating. Because that first version was laid out much like the current one, Fred had had to determine a centre point, just as we had to.

"I have to admit, I never expected to find his original mark. But when all the seats and the floor had been torn out, and we were down to the lower level, lo and behold, there beneath us was a slab of concrete with a mark cut into it. And of course we realized right away that it was Fred's original centre point for the tent from 1953.

"The more amazing thing yet was that when we calculated our own centre point it was within an inch of his. The theatre today stands precisely where the tent stood more than 60 years ago.

"For me, the beauty of that sort of discovery is partly just the pleasure of discovery. But it's also what it represents – in this case, the historical continuity of a great theatre and, beyond that, the continuity of what surveyors do, of how they contribute, generation after generation, to the larger creative enterprise."

Peter Moreton earned his commission as an Ontario Land Surveyor in 1972, and a Canada Lands Surveyor in 1982. He is among other things a lover of music and over the years has attended numerous plays and musicals in the Festival Theatre.

He is seen here through the lens of a Trimble S3 Robotic Total Station.

Clockwise from top: 1) The original Festival tent in 1953. 2) The circular floor plan of the Festival Theatre. 3) The Festival Theatre at night. 4) The thrust stage at the precise centre of the theatre.

1

4

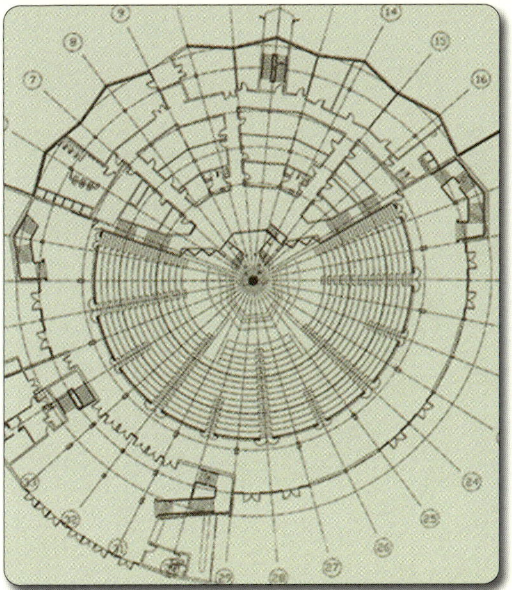

2

3

Futurama – City of Tomorrow, *oil on canvas, by Moma Markovich, 1968.*

Government of Ontario Art Collection, Archives of Ontario 101207.

CHAPTER SIX

BOOM AND CONSUME

THE GREAT ECONOMIC BOOM that followed World War II consumed Ontario to about the same degree as Ontario consumed the economic boom. Between 1945 and 1955, the province's population (some 4 million during the war) increased by 1.5 million, including more than 800,000 immigrants. Overwhelmingly, the newcomers flocked to the cities, where, like the rest of Ontario's war-weary inhabitants, they needed an expansive new world in which to live, to work, to study, to travel, *to consume* – which is to say to buy, and buy some more, to bring it all home, as no society anywhere, at any time in history, had ever consumed. They bought houses and cars and cottages; refrigerators and dining room sets; radios and televisions. And more televisions. All of which required manufacturing sites, showrooms, board rooms; required distribution, which in turn required roads and rail lines and airports. Required surveyors as never before.

Within the surveying profession, which had been staggered by the twin blows of the Depression and war, the number of articling students rose from a low of three in 1944, to twenty-two in 1945, to sixty-one in 1946. AOLS membership had bottomed out in 1944 when just 201 commissioned land surveyors were represented by the organization. By 1952, the membership stood at three hundred; by 1956, four hundred; and by 1960, more than five hundred.

From one end of the province to the other, this energetic new body of surveyors was subdividing land, plotting highways, planning hydro dams and power lines; surveying timber stands and mining claims, including uranium discoveries around Elliot Lake. For the first time in Ontario's history, whole towns were being created at a swoop. It is neatly symmetrical that the town of Elliot Lake was invented holus-bolus to take uranium out of the ground, while the town of Deep River (add water and stir) was built to support the nuclear reactors that consumed that uranium.

YORKDALE
Shopping
Centre

Opposite page, top: Gleeful consumption and the construction of highways and housing drove much of Ontario's expansive suburban economy during the 1950s and '60s. It is said that for every acre of selling space at the average mall built during that period, five acres of asphalt were laid down for parking. Yorkdale Mall, one of the world's first and most sumptuous indoor shopping centres, was clearly no exception. Bottom: Eaton's Vista restaurant, at Yorkdale, captured the expansive and jazzy spirit of the times. Right: The construction of the Trans-Canada Pipeline across Ontario during the late 1950s employed dozens of Ontario land surveyors.

The construction of the Trans-Canada Highway was expansively (and expensively) in sync with the spirit of an age in which big spending seemed a necessary response to the privations of the Depression and war; frugality was for chumps. On parts of the route north of Lake Superior, the cost of building a mile of hardtop – what with the rock cuts and bog-filling and bridges – was equal to the cost of building fifty miles of highway on the prairie.

If the frenetic activity around road construction (the grinding of the landscape, the movement of vehicles and people and goods) seemed a kind of allegory for the bold new cultural reality, so too did the construction of the Trans-Canada Pipeline, one of the biggest surveying and engineering projects in Canadian history. In Ontario, much of the surveying work on the pipeline was contracted to the well-regarded Toronto firm of Marshall Macklin Monaghan. Begun in 1956, the pipeline entered Ontario west of Kenora and exited the province south of Ottawa, some 2000 kilometres to the east. More than 40 Ontario land surveyors and survey technicians worked on the pipeline's alignment and routing, as well as on the dozens of pumping stations that moved (and still move) the gas across Ontario.

Meanwhile, in Ontario's cities, cadastral surveyors worked ten-hour days, more or less throughout the 1950s, to stake out and define the rapidly expanding suburbs. When they had successfully established streets and lot lines, their focus shifted to parks and shopping malls, to arenas and schools – and of course to the ever-teeming parking lots that were increasingly the *sine qua non* of suburban development. By 1955, every new public building and facility in the suburbs came aproned by several acres of Stygian black asphalt, every house by a miniature equivalent.

In the Don Valley, northeast of Toronto, one of the world's most advanced city-planning endeavours was initiated in 1952, once again in consultation with the Toronto engineering and survey firm, Marshall, Macklin, Monaghan. What until then had been a tract of twenty 100-acre farms owned by the Canadian billionaire, E. P. Taylor, was, during a period of a dozen years, transformed into the futuristic suburb of Don Mills.

Every feature of this architecturally fastidious community was planned to the near-Utopian standards of science fiction. The houses were constructed of what a prospectus called "uniformly attractive materials" – often in "bun-

Dr. Alec McEwen, OLS #784, enjoyed one of the most extraordinary careers in the history of Canadian surveying, working in 25 countries during the latter half of the 20th century, as well as half a dozen U.S. states and Canadian provinces. For several years during the late 1950s, he was chief surveyor for the Toronto Transit Commission, and from 1976 to 1990 headed the Canada-U.S. International Boundary Commission in Ottawa and Washington. Along the way, he earned five degrees, including three in law, and held a professorship in engineering at the University of Calgary. He is seen here operating a microwave distance-measuring MRA3 tellurometer, one of the first pieces of revolutionary post-war surveying equipment.

galow" or "ranch-style" design – the lot of them mixed with "low-density" apartment buildings and row houses that would draw a cross-section of the social strata of the city. The schools and municipal buildings were designed in the minimalist style of the Bauhaus, while the streets featured curves, T-intersections and cul-de-sacs, in order to slow down car traffic (in other words, to slow down vehicles whose speed and efficiency was intrinsic to the existence of the suburb).

Pedestrians were separated from cars through the use of foot paths, so that people could walk to school or to the bus stop or baseball diamond – or even to work, given that the state-of-the-art subdivision included offices and warehouses and a variety of structures in which light-manufacturing could go on. The message was, life and work are one; planners were determined that Don Mills not become a bedroom community for some other part of Toronto. In an era long before people connected the health of the natural environment with the health of a human community – or, for that matter, before ecology was even a familiar word – a "green belt," consisting of several small parks and portions of the local ravine, graced (as well as oxygenated) the new suburb. Part of the green space became a ski hill in winter – the first urban skiing facility in North America.

The population, meanwhile, was not the random aggregation of people who might settle on a given patch of the planet but, rather, 28,000 exemplars of what a perfect suburb's inhabitants ought to look like: moms and dads and children, in ideal proportions, all law-abiding, all faithful in their commitment to the future – seven thousand of them living in each of four "neighbourhoods," as the suburb's quadrants were called.

It hardly needs saying that the irreproachable centrepiece of the great suburban experiment – the place of harmony at the crossroads of the quadrants – was neither a civic square nor park nor reflection pool, nor li-

Houses in the Toronto suburb of Don Mills were generally sleek, low-profile affairs that featured expansive driveways and, often, two-car garages or carports. Below: A mid-1960s aerial view of the famous modern suburb.

brary nor art gallery nor public garden, but of course a shopping mall, a place of consumption, the Don Mills Centre, with a parking lot big enough to handle several thousand of the suburb's shiny new Impalas and Parisiennes and Galaxies.

During the early stages of its development, Don Mills was a *de facto* organizing principle for land surveyors – a host of them, busily laying out roads and lot lines and drainage slopes; and of course underground infrastructure, which in this case brought together everything from electrical and telephone wiring to water pipes, sewer pipes and storm drains.

By 1960, the original 2,000 acres – land on which E. P. Taylor had once planned to build a brewery – had become a showplace for what was both best and worst in suburban planning, as well as a model for subdivisions across North America.

With faith in oil and automobiles running on a decidedly full tank during the early 1960s, no one questioned the degree to which a suburb such as

The most distinctive building in Don Mills during the 1960s, the white Don Mills Curling Club, seen above, was converted to a bingo hall during the 1970s and torn down in 1986 to make way for a strip mall, which itself has since been torn down in favour of a condominium tower.

Below right: Concrete spaghetti – showing the Eastward view of Highway 401 as it passed north of Yorkdale Shopping Centre in 1966.

The Little Pic River bridge, just north of Lake Superior, was one of 25 such bridges required during the late 1950s to complete the 300-kilometre stretch of the Trans-Canada Highway between the Agawa River, west of Sault Ste. Marie, and the town of Marathon, 300 kilometres east of Thunder Bay.

Little Pic River Bridge, *oil on board, undated, by Moma Markovich; Government of Ontario Art Collection, Archives of Ontario 101218.*

Don Mills created a kind of divinity – not to mention a crutch – of the automobile. "You couldn't live there without one (or two, or even three), unless you and your family wanted to live *only* there," said David Kirk, a long-time resident of Don Mills, in the spring of 2015. "You went shopping in the car, went to the restaurant, the cottage, the ski hill, the dentist; you took the kids to their hockey games and piano lessons in the car." If you worked outside of Don Mills, you drove to work (in fact, you probably drove if you worked within Don Mills). Kirk tells an apocryphal story about the east-Toronto surveyor who during the mid-1950s drew up a plan for a quarter mile or so of suburban Lawrence Avenue, factoring in generous sidewalks on both sides of the street; and of the city supervisor who brought the plan to a meeting and demanded to know: "Why on both sides? Nobody walks anywhere anymore!"

Thirty years would pass before surveyors, engineers and city planners would begin thinking about communities that were *not* so dependent on cars and on finite resources such as oil. At the time, however (as in the era of "Good Roads" Campbell), highways were being rolled out like measuring tapes, while existing roads were being widened or improved or extended. Highway 401, once a futuristic chimera, was by this time a functionally banal reality that could get a driver from Cornwall to Windsor in seven hours. Highway 400 (with a little help from #11) could get a driver from Toronto to his or her cottage in Muskoka in an hour and 45 minutes. The latter was a drive that 20 years earlier had taken as long as four hours over back roads, and 20 years prior to that was generally accomplished by train from Toronto to Gravenhurst, and from there by steam-ship up the Muskoka Lakes, the lot of it consuming the better part of a day.

Ivan Wallace • And the Magic Would Begin

In 1961, the surveying firm of Horton and Wallace, in Whitby, imported from California what was known as a second-generation computer. The reconditioned machine was as big as an office desk and, according to Ivan Wallace, a partner in the firm, weighed "as much as three pianos." It cost $15,000 – approximately the price of a 3-bedroom house at the time. For a smallish firm, the computer was a significant investment. It was also, Ivan believes, the first computer used by land surveyors in Ontario.

"I'm talking about a world in which microchips didn't yet exist," Ivan said in an interview in February, 2015. "This massive thing operated with radio tubes, in effect heat producers, which created a problem because the rotating drum inside the computer had very low tolerances for expansion and contraction and needed a cool ambient temperature. In the early 1960s, central air-conditioning wasn't common, so we'd no sooner taken the thing off the flat-bed that delivered it than we were obliged to install an air conditioner in the window."

The computer also required what Ivan describes as "absolute dedicated electricity." To achieve this, the firm persuaded the local Public Utilities Commission to install a hydro pole and transformer directly adjacent to the office, which cut down on fluctuations in power.

"I didn't know the term 'idiot savant' at the time," said Ivan, "but that's what my partner Graydon Horton was – an eccentric genius. Without guidance of any sort, he figured out how to write programming so that this computer could work with curves, get astronomical bearings from sun observations, do all sorts of calculations – the sort of things that today's surveyors take for granted. We did have electronic calculators in the office prior to this – but for the field staff it was logarithms or go home."

Among the computer's drawbacks was that it didn't have a monitor, meaning that the operator couldn't see what he or she was cal-

Ivan is seen here in his office during the provincial election of 1977, in which he ran as a Liberal candidate. His niece, Joanne Bond, who worked in the office as a draughtsperson, is seated at the desk.

The second-generation computer depicted at left required everything but its own nursemaid when it arrived from California at the offices of Horton and Wallace, in Whitby, in 1961.

culating. Nor did it have a keyboard for entering information. Data was entered, rather, using a teletype machine that punched holes in inch-and-a-half-wide paper.

"I'd punch all the information into the tape, which would take maybe ten minutes," said Ivan, "Then I'd examine the tape, patch up any errors, and when it was all ready to go would feed it into the machine. And the magic would begin. Or not begin if you'd made a mistake. In which case you started over.... We operated that thing for half a dozen years – constantly learning new efficiencies and improving our field techniques. Once we got up to speed, we could do builders' surveys in a single morning – tie in a dozen or more structures and produce the plans the same afternoon. This was revolutionary at the time."

By the late 1960s, Toronto-area surveyors had begun to use computer terminals connected by phone to a main frame, operated by IBM, in the city. "It was a good system," Ivan said. "The problem was that from Whitby, if you were going to do any serious calculation, the long distance charges amounted to a small fortune. So Horton and I pretty much proceeded as we always had – on our own. I contributed my practical knowledge, and 'Hort' contributed the brilliance that allowed us to develop our software – a very fine marriage of practicality and abstract intelligence. And that is how in our Whitby office, during the early 1960s, we brought computerized surveying to Ontario."

Graydon Horton

Ivan Wallace was commissioned as a Ontario Land Surveyor in 1960 and today operates Ivan B. Wallace Surveyors Ltd. in Bowmanville. Renowned for his community involvement, Ivan served for 24 years as a trustee of the Durham Region School Board; volunteered for 24 years with the John Howard Society, including six years as its national president; served for six years on the Ontario Board of Parole and for twelve years as Chairman of the Appeal Board for Employment Insurance. For his efforts in the community, he received the Apostolic Blessing of Pope John Paul II in 1982 and was awarded the Queen's Silver Jubilee Medal in 1977.

International cargo vessels using the St. Lawrence Seaway are built to a maximum length of 230 metres and a width of 24 metres, allowing them to fit with just a hair's breadth of clearance into the system's 15 standardized locks.

The Moses-Saunders hydroelectric dam at Cornwall produces 1,045 megawatts of electricity and elevates the water level upstream to navigable depth for ships using the Seaway.

As in the Book of Genesis, where Seth begat Enos who begat Cainan: 400 begat 401, which begat 427, which begat 409, which begat 410 and 404 and 407; and 416 and 417 – all of which, in their way, were descendants of the infinite skein of smaller highways, such as 2 and 5 and 6... and 8 and 11 and 17, the lot of which begat the equally indefatigable construction of bridges and culverts and rest stops; and lighting systems and signs and interchanges; and a concern for safety and drainage: all part of the complex functionality of a scheme that brought together people, wheels, roads, cargoes and destinations.

And, once again, surveyors.

The thought may never occur to those who drive Ontario's highways, but not a metre of any of those highways, not a millimetre, has ever been planned, plotted or constructed without the participation of a commissioned land surveyor – sometimes many surveyors, in many capacities.

Not surprisingly, during the dozen years that followed the war, scores of surveyors were articled directly to the legendary Harold Howden, Ontario's Chief Inspector of Surveys, in charge of thousands of miles of the province's highways and secondary paved roads.

"I seem to remember that during the time I was articling under Harold," says Jim Hill, "he spoke to me once, in passing. All my real education came from other people. But because he was

the top surveyor in the organization, he was the guy whose name appeared on our articles."

Many of Howden's charges eventually took their skills to other areas of surveying, but none ever forgot that the traditional width of most drivable roads in Ontario, including the shoulders and right-of-way, was 66 feet – precisely the length of an old-fashioned surveyor's chain.

HUNDREDS, IF NOT THOUSANDS, of post-war projects and developments came to embody both the meaning and indispensability of good surveying: major airports, suburban development, freeways, subways, skyscrapers, pipelines, bridges, tunnels, water diversions, vast hydro and nuclear undertakings. However, if one project more than any other came to symbolize Canada's industrial and economic development during the 1950s and '60s, as well as the importance of intelligent and imaginative surveying, it was the construction of the St. Lawrence Seaway and hydro project, one of the greatest engineering feats in history.

The purpose of the Seaway was to allow ocean-going cargo vessels to travel up the St. Lawrence River into the heart of North America via the Great Lakes – to reach deep-water ports such as Toronto, Cleveland, Detroit, Chicago, Milwaukee and Thunder Bay, which are all, in effect, seaports, some as far as 3,000 kilometres from any sea. Prior to 1959, ocean vessels with cargoes bound for, say, Chicago or Milwaukee, had to dock at Montreal so that their cargoes could be transferred onto much smaller lake vessels that would fit through the narrow canals that bypassed the river's many rapids and shallows.

Ocean access via the St. Lawrence had been proposed as early as 1890, with varying levels of support from both sides of the border. But by 1945, the U.S. Congress had soured on the idea, preferring to support America's Atlantic ports and the inland railways that would be undone by the Seaway. The result was that in 1951 the Canadian government, impatient with the Americans, declared that it would build the Seaway on its own, making it an entirely Canadian endeavour.

However, in 1954, unwilling to be shut out of what increasingly promised to be an unprecedented boon to the North American economy, the U.S., under President Dwight Eisenhower, recommitted to the mega-pro-

Flooding upstream from the Moses-Saunders hydroelectric dam at Cornwall in 1958 inundated nine historic villages – the so-called "lost villages" – in Cornwall and Osnabruck townships. In some of those ghost villages, such as the flooded remains of Aultsville (above), the remnants of streets and building foundations can still be seen beneath the water. Those remnants were photographed from the air in a special project by Louis Helbig in 2013.

ject, and engineers, surveyors and construction firms on both sides of the border got down to business. (It is worth noting that, despite apparent U.S. reluctance during the late 1940s and early 1950s, surveyors from the U.S. Corps of Engineers had, by 1954, already produced some 13,000 pages of plans for the project.)

One of the key features of it all would be the immense Moses-Saunders Dam at Cornwall, which would produce hydro power for both Ontario and New York State. The dam would also raise the level of the river upstream, so that it would be navigable for ocean-going ships, or "salties" as they are known on the Seaway.

The waterway would consist of 15 locks between the Port of Montreal and Sault Ste. Marie, plus hundreds of kilometres of dredged channels in various parts of the St. Lawrence and Great Lakes.

From a Canadian surveyor's perspective, the immense project got off to a resonant start when Toronto surveyor, Norman Wilson, who at the time was the president of the AOLS, made initial surveys intended to discover the projected extent of the flooding that would occur with the construction of the Moses-Saunders Dam.

In all, 6,500 people on numerous farms and in nine villages – the so-called "lost villages" – were forced to move because of the flooding. Towns such as Morrisburg, Long Sault and Ingleside, all of them in far eastern Ontario, were constructed to receive those whose farms or town lots were about to disappear under water. Hundreds of houses, some nearly a century old, were moved to the new communities – moved so carefully that their owners were encouraged simply to leave furniture, décor, lamps, dishes, and the like, where they were, unsecured, while the moves took place. A number of cemeteries doomed to inundation were relocated, grave by grave.

The surveying work lasted more than a decade and extended to everything from lock and dam construction to hydrographic issues tied to changes in the riverbed, to the creation of shipping channels, the construction of

Buildings from villages about to be flooded by the Seaway in 1958 were removed from their foundations, picked up intact, and hauled to the newly created towns of Ingleside and Long Sault. The town of Iroquois and parts of the town of Morrisburg were moved to higher ground and still bear their original names.

docking and port facilities... to bridges and roads and spur lines on the railway, to laying out new villages, creating new cemeteries, and of course to the determination of where exactly the new shorelines would be when the water rose.

Like most vast infrastructure projects, the making of the Seaway took on a kind of socio-moral weight – for God, for country, for humanity – but on a simpler level it was about commoner ambitions: the movement of goods, of resources, of fuel, of whatever it took to keep the power coming, the factories producing, the buying and selling going non-stop, forever, night and day.

In 1958, the inundation came, and in 1959, the first ocean vessels, monsters by comparison to any other ships Ontarians had ever seen on the Great Lakes, made their way through the locks into the upper lakes.

The late Hazen Meldrum was one of hundreds of surveyors who had flocked to the project from various parts of Canada and the world. "It was toward the end of the construction of the Seaway, maybe in 1958, that he set up his own survey office in Cornwall," said Hazen's daughter, Julia Meldrum Smith, in May 2015. "I always knew he'd had a role with Hydro in the building of the Seaway, which is what had brought him up here from Nova Scotia. Many years later, after I'd become a surveyor myself, I said to him, 'Dad, why did you suddenly stop with Hydro and set up on your own?' He said, 'Because they let me go! The Seaway was done!' It had never occurred to me that he'd lost his job!"

Meanwhile, jobs for surveyors were almost anywhere one cared to look during the 1950s and '60s: building the suburbs, building skyscrapers, building more highways, more bridges, more airports; mapping the north; mapping the floors of the Great Lakes; establishing the Distant Early Warning system or DEW Line. The last of these, another joint Canada-U.S. project, created a 10,000-kilometre necklace of radar stations across the broad shoulders of the far north, some 250 kilometres above the Arctic Circle. The Cold War was at its height, and the purpose of the 63 stations was to detect attacks by Soviet bombers coming across the polar region, carrying nuclear weapons. Understandably, the construction of the line – its radar installations, its living spaces, its communications and warehousing – was a nightmarish job, conducted at temper-

Surveyors, engineers and construction crews worked quickly after World War II to carve two "instant" towns, Elliot Lake and Deep River, out of the Ontario wilderness. Incorporated in 1955, Elliot Lake was home to the province's uranium-mining industry for some 40 years, while Deep River, incorporated during the mid-1940s, provided a semi-utopian home for the scientists and engineers brought to the town to build and run the country's first nuclear reactors at nearby Chalk River. During the mid-1950s, Mike Maughan, a young surveyor with Marshall Macklin Monaghan, spent a year living in a tiny trailer amidst felled trees and blasted rock, in charge of four crews that worked ten hours a day, six days a week, laying out the townsite that would become Elliot Lake (seen above in early development).

atures that were at times so low they were not even measurable with conventional thermometers. The project required the labour of 25,000 men and women and began, as always, with surveyors, men who flew over some of the harshest terrain on earth and selected locations for the 63 outposts and later dealt with site plans and technological issues and infrastructure. The surveyor of earliest record on the project was John Anderson-Thompson, a Dominion Land Surveyor located in Yellowknife, who selected the station sites from the air. However, later phases of the project drew on the skills of numerous others, including Ontario practitioners such as the future Erindale professor, Bob Gunn.

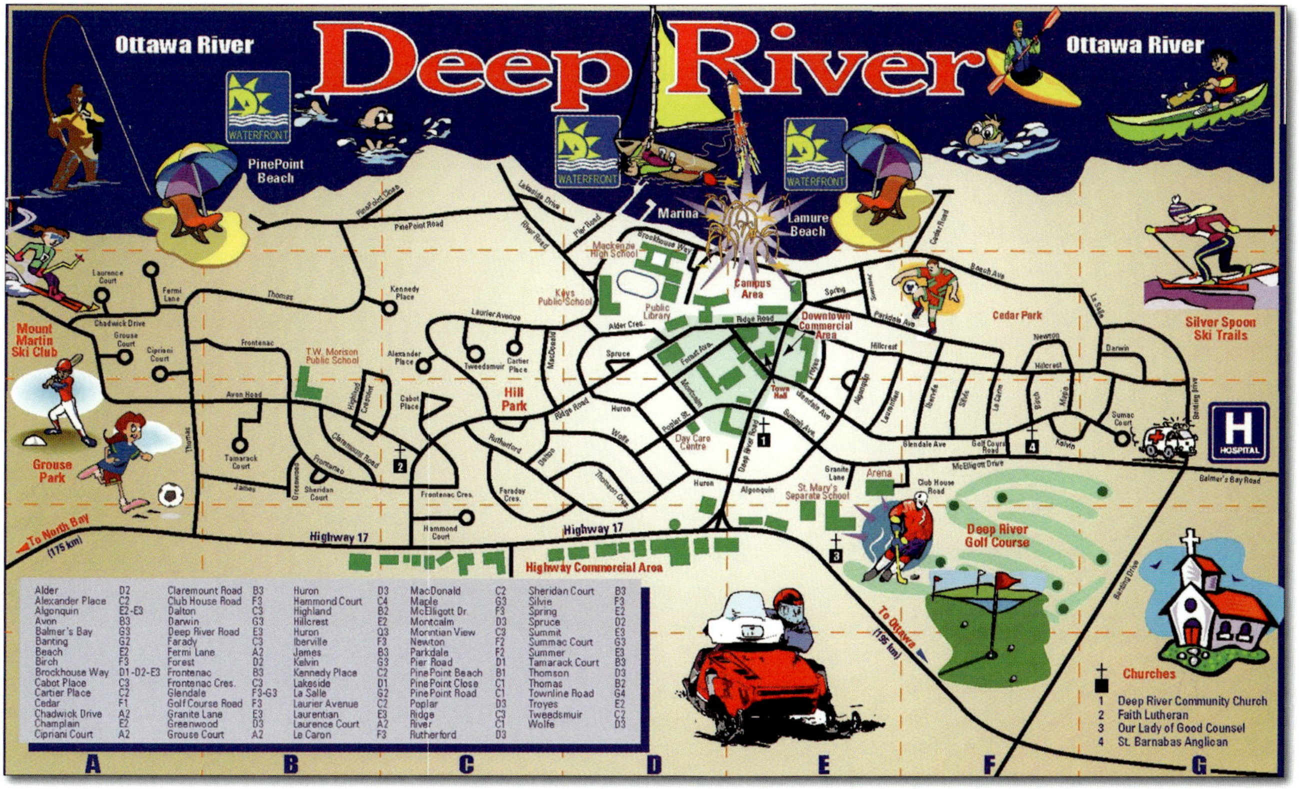

IT MIGHT SEEM THE HEIGHT OF PARADOX to a present-day observer, but as Ontario's surveying projects got bigger and more complex, the general quality of surveying in the province was believed to be deteriorating. Blake Erwin, the president of the AOLS in 1958, attributed the widely perceived drop in standards to the great demand for surveyors coupled with low fees, which he believed tempted some surveyors to cut corners. Others in the Association, including the esteemed Fred Pearce, who would eventually be known as "Mr. Survey Education," believed the problem went deeper and would not be solved until the standards of education for surveyors were raised, and the articling system was updated and refined. "Bad practices," he claimed, "are a result of well-intentioned surveyors who have fallen behind in their methods and practices, and are passing along their limitations to their apprentices." The difficulty was aggravated, he said, by what he called "accelerating advances in technology" – advances presumably beyond the scope of some surveyors.

If better education was needed, so too was a more progressive surveyors' association, one that would promote professional development and a better public image, and would apply a rigorous review process to the work of the roughly 500 surveyors whom it represented and regulated. There was also, it seemed, a need for tighter disciplinary procedures against delinquent or careless surveyors and a more accessible complaints procedure for the public.

Whatever the causes of the deterioration, the remedy – or at least the seed of a remedy – was addressed in short order by some of the most enlightened surveyors and educators in Ontario. For one thing, legislated standards for surveying were introduced under the Land Titles Act during the early 1960s. Those standards were written by David Lambden, the first Examiner of Surveys under the leadership of Marsh Magwood, the esteemed Director of Titles.

By 1962, the University of Toronto had initiated what it called a "survey specialty" as part of the civil engineering program at its St. George or "downtown" campus. Those who

During the years after the war, Ontario's expanding southern cities were a kind of siren song for young surveyors. But some heard other music. The photo above, taken in 1954, shows (left to to right) Jack Monteith (soon to be OLS #1053), with Bob Glenn, Peter Buedon and Larry Monaghan. The quartet was conducting an 83-mile baseline survey in the wilds near North Spirit Lake, some 600 kilometres northwest of Thunder Bay. At the time, Monteith was articling to the London surveyor, Colonel Sam Archibald. On another day in the wilderness that same year, young Monteith observed the Colonel hanging his underwear to dry over a cook fire, and immortalized the occasion on film.

Susan MacGregor • Five Thoughts from the Purveyor of the Surveyor

1 **"When I was named Surveyor General in 2010,** it took me a couple of years to figure out exactly what my job was, because it has this ever-expanding orbit of work around it. We do so much: parks, mines, fisheries, boundaries, development control, private property issues. At the same time, we're involved in numerous First Nations land claims in Ontario – we assist with the negotiation process, as well as the resolution. It's great that our services are in such demand, but the result is that the opportunities to think *big*, to think strategically into the future, are really limited. I try to do that in the evenings and on weekends, because I believe we as surveyors contribute something really important to society, and that we must continue to deliver it a nd not fail. I've tried to raise awareness of this and admit it's been a struggle to get the message across."

2 **"Think of the way Ontario developed.** We had these post-war infrastructure projects to put people back into the workforce when they returned from overseas. Today, we need to know where the vulnerabilities are in these aging infrastructure systems, so that in the context of increasingly severe climate or weather events, we can make sure that when we flip the switch the light goes on, or that the natural gas continues to flow. Flood mapping, for example, is going to be more and more important and will require a big investment. Our challenge moving forward is about managing what used to be hundred-year storms or fire events that are now ten-year events."

3 **"In the 1980s surveyors dreamed big.** It was through their efforts, for example, that the Ontario Base Map Program was created, meaning that Ontario was mapped to a much higher quality than ever before. Later, in the early 1990s, the paper maps that were created became the MNR's foundational GIS system, what we call NRVIS (the Natural Resources Values Information System). Creating this depended on the ability of surveyors with foresight to get in front of the decision makers and explain what was necessary and what was possible. These days, we have GPS and Google Maps, and all sorts of other marvelous tools, but very few people understand the complexity of these marvels or the investment needed to create them, which to a large degree has involved surveyors."

4 **"Failure is not an option** for surveyors. As a society, we just expect them to do what they're supposed to do and do it well, even though the culture lacks any broad understanding of what that is and why it's important. The truth is that without accurate surveying, even our banking system would be in jeopardy. Our capacity to define the limits of what we own underpins our ability to leverage land by way of a mortgage. It's foundational. Together the surveyor and lawyer create certainty and security for land ownership and, with that, enable business investment and peaceful neighbourhoods. If this is all done well, we promote social and economic investment in Canada."

5 **"At the moment,** we don't even have a university program specifically for training surveyors. And there are just sixty or so Ontario Land Surveyors under the age of fifty. If there were a direct path for people to go from grade twelve to being surveyors, a fair number of our challenges would begin to sort themselves out. I do feel a weight of obligation to help create a university surveying program, but I also feel that the office of the Surveyor General can't possibly do it on its own. So my focus is on raising a broader awareness of surveying's contributions and social responsibilities. I guess in a sense that makes me and my office the purveyor of the surveyor."

John Goltz • Polaris at Any Time of Day

John Goltz, OLS #1185, received his commission as a surveyor in 1965. He works with McIntosh Perry Surveying Inc. in Perth.

"**As a young fellow, in the 1960s, I made a decision** to become a teacher. I went to Queen's and got a degree. Along the way I started working on survey crews in the summer. That was my downfall, because gradually I fell in love with the wonderful combination of mathematics and law and geography that is the foundation of land surveying. Best of all, in those days, I grew to love astronomy, making sun and star observations. The calculations weren't easy but they weren't particularly difficult either – you knew where your star was, you knew where the centre of the earth was, you knew where you were standing on the earth and where your particular meridian was. Using trigonometry, you could work out this great galactic triangle that gave you a precise bearing on your own tiny part of the planet.... I became so adept at this, so fascinated by the stars as they pertained to surveying, that eventually I could find the star Polaris even in the middle of the day. If you knew where to look, knew the altitude and the azimuth, you could find it.... Nowadays, all of this is gone from surveying, of course, and I don't lament it. I suspect that for young surveyors the technologies that have replaced it – LiDAR and GPS and UAVs, and so on – are probably just as big an attraction as astronomy was for me."

George Annis, left, and John Goltz, at the Annis, O'Sullivan & Goltz northern office, Frobisher Bay NWT, 1975.

Lost in the Defence of Canada...

Found by Jim Purcell

Many surveyors have known the surprise and pleasure of discovering a rare object while digging for evidence of a vanished boundary or monument. In most cases, it is a humble item – say, an old cedar monument preserved in the anaerobic conditions of a bog; or a hand-forged nail, bearing the initials of a long-dead surveyor; or a chunk or two of broken crockery buried to mark a property corner.

In 1867, a steamboat entrepreneur named Daniel Cowley found what is said to be Champlain's astrolabe, in a swamp near Cobden, Ontario – a treasure lost by the famous explorer as he made his way up the Ottawa River in 1613, and now housed in the Canadian Museum of Civilization.

Among modern finds, few match the discovery made by Jim Purcell, OLS, on a quiet afternoon, in 1957, as he surveyed a property on Hamilton Mountain, in Flamborough Township, while articled to Harry Williams at Ontario Hydro. "I was look ing for evidence at a fence post," he says, "when my shovel clinked."

Within seconds, Jim had unearthed an antique gold pocket watch, intact except for a broken hand and a lost winding key. "Oh, it was a beautiful thing," he says, "– yellow, white and blue gold; very delicate. It weighed a ton."

Jim took the timepiece (the face of which is seen here) to a watchmaker, who got it working, discovering in the process that it had been made by the famous English watch manufacturer, J.M. Tobias, of Liverpool, in 1811. "The most amazing thing about it," says Jim, "is that it runs on a system of tiny gold chains, rather than the customary gears." In 2013, Jim took the the watch to Ashland Appraisers, in Sarasota, Florida, where it was evaluated it at upwards of US $4500. "My belief," says Jim, "is that it was owned by one of General Brock's soldiers, probably an officer, and that he lost it defending Canada against the Americans."

operated the program included a trio of engineers – Bob Gunn, Oscar Marshall and Basil Haynes – who would eventually have roles in founding the larger and more specific surveying program at Erindale. More than a stop-gap, however, the original program was a solid on-ramp to what was coming. One of its graduates, the surveyor Tom Bunker, of Gravenhurst, said in January 2016, that he believed "the instruction offered in that short lived program (1962-1970) matched any given later at Erindale."

Nearly a decade would pass before the AOLS set a radical new course by demanding that all new land surveyors in Ontario must acquire a university degree. Meanwhile, during the late 1960s, an education committee was created by the Association to prepare the way for what would eventually be established. Central to the committee were men of vision and energy such as the aforementioned Fred Pearce, a much-admired Stratford surveyor, who did not himself have a university education but who understood the stakes and became chairman of the group; and the indefatigable Jack Barnes, whose progressive attitudes had made the J.D. Barnes firm in Toronto the largest and one of the most progressive survey firms in the province; and the engineering professor, Oscar Marshall (of Marshall Macklin Monaghan), who for years had advocated a university program devoted exclusively to the art and science of surveying.

The functioning of the committee entailed an extended negotiaton with the province, the Surveyor General's office, and of course the U of T's Erindale College, which, under the principalship of the renowned geophysicist John Tuzo Wilson, was perceived to be a place where the survey program could and would (and ultimately did) thrive.

Next came the hiring of a director, the genial and talented engineer and photogrammetrist, Gordon Gracie. "We all knew Gordon," recalled Bob Gunn in late 2015. "I knew him from World War II when we took under-age training together at No. 1 Manning

Erindale professor Bob Gunn (right), with students (right to left) Andy Cameron, Rob McQuaig and Don Cross, at Gull Lake Survey Camp during the summer of 1969. The students were enrolled in the survey option of U of T's engineering program.

Depot, in Toronto, so that we could enter the Air Force when we turned eighteen. Oscar Marshall had once hired him to work on the Trans-Canada Pipeline, for which Marshall Macklin Monaghan had the surveying contract in Ontario. So he knew him and respected him. And a number of us had continued our association with him during his years of research and teaching at the University of Illinois."

"The first faculty member we hired was Bob Gunn," said Gracie in May 2015. "Until then, he'd been with the civil engineering faculty at U of T's downtown campus." Jack Young, another heralded Erindale professor, called Gunn "an excellent all-around technical surveyor and a fine teacher," adding that when Gunn got a citation a few years ago at the annual general meeting, the chair of the event asked everybody who had attended Gunn's lectures to stand. "There must have been 400 people on their feet, clapping; this was out of the 500 in the room. So he had quite an affect."

Gracie's second recruit was the program's brilliant and eclectic survey law professor, David Lambden, who was respected by his students perhaps more for his erudition and knowledge

The castellated exterior of Erindale College during the early years of the Survey Science program. The building, construction of which began in 1966, was one of Canadian architect Raymond Moriyama's first major commissions.

Andrea Tieman, OLS #1635, receiving her convocation certificate in 1988 from Michael O'Sullivan, who in 1994 would become Surveyor General of Canada. Andrea's future husband and fellow graduate, John Galejs, can be seen behind her to the left.

What Do the Girls Think?

"As with many young people coming out of high school, I had no idea what I wanted to do with my life way back in 1979. I loved math, but my marks generally weren't great, so that my only university acceptance was at the Erindale campus of the University of Toronto. I enrolled in business, then switched to economics. But I was never happy with those choices. One day, someone told me they had a friend with an economics degree who worked in a bank, and I certainly had no interest in doing that.

"So, I began looking around. I'd seen students wandering the campus with theodolites, measuring distances and buildings, and so on. I was intrigued by this and on a whim the following year decided to sign up for a single first-year course in Survey Science – taught by Jack Young, who, as it turned out, swore me in as an Ontario Land Surveyor six years later and presented me with my 25-year-pin a quarter of a century after that.

"Pretty much right away, I knew I'd made the right choice. The course involved math and working outside, and there was a genuine sense of camaraderie among the students and professors. The following year, I committed completely to the program. I found the technical courses enjoyable and challenging, but thinking back I realize too what an impact David Lambden had in sparking my interest in boundary law and history. I was terrified of him. I wasn't one to speak up in class, and he would often ask, "What do the girls think?" There were only two of us.

"I was sworn in as an Ontario Land Surveyor in 1988, just the eighth female OLS. These days, attending the AGM is the highlight of my year. In seeing the people I went to school with and those I've worked with over the years, I feel the same sense of camaraderie that I felt in that first Survey Science class back in 1982. I always attend the meeting's opening ceremonies, and particularly appreciate the moment of silence for Ontario surveyors who've passed on during the year. It means a lot to me to be in a profession that honours its past, and those who made it what it is."

– Andrea Tieman

A gathering of newly commissioned OLSs and early grads of the Erindale Survey Science program (with honoured guests) at the 1979 AGM in the Westin Hotel, Ottawa. 1. Past president Mike Maughan, 2. Executive Director Lorraine Setterington, 3. President Don Endleman. Class members: 4. Ken Murrary, 5. Paul Church, 6. Doug Simmonds, 7. John Gutri, 8. Kim Clement, 9. Stefan Eperjesi, 10. Ron Stewart, 11. Eric Van Soeren, 12. Jim MacIntosh, 13. Harry Gustafsson, 14. Dave Urso, 15. Jim Statham, 16. Izaak de Rijcke, 17. Alban Rattos, 18. Brian Williams, 19. John Ward, 20. Blair Culham, 21. Blain Martin, 22. Marcel Renaud, 23. Dave Woodland, 24. Maureen Mountjoy.

than for his oft-elusive teaching manner. "Oh, he could be frustrating," said Jack Young. "Before I started teaching, I actually took his course. He'd draw a section of land up on the board, and he'd show the students where the survey posts had been found, and then he'd present some complicated problem, involving where one of the property lines should be located. He'd discuss the various options with the class, and then without ever telling them where *he* thought the line should go, he'd say, 'Let's go for coffee!' And when he got back, he'd immediately erase the blackboard and move on, leaving them guessing. Don't get me wrong, I have a very high regard for David and his knowledge of survey law and his ability to survey, but I'm sure those students would have appreciated some guidance as to how he weighed the evidence and where *he* would have put the line."

In the words of Gordon Gracie, "There were always controversies. Universities seem to engender them. But there was also a feeling that these were

privileged years – a grand opportunity. We had a genuine feeling that we were building the profession, raising its arc – it was no longer going to be any sort of old boys' club or avocation."

Beyond educational standards, Erindale is generally perceived to have altered the way surveyors interacted in the province. "In the 1950s and '60s," said the surveyor Gord Good in February 2015, "there was a flood of new surveyors to meet the demands of the post-war boom. But very few of them really knew one another – they hadn't been to school together, didn't really have any basis for cooperation or camaraderie." Good's thesis is that when the boom began to fade, those surveyors tended to grow competitive about available work, to build what Good calls "an exclusionary zone" around their home turf. "When the Erindale surveyors started graduating," he said, "they'd been at school together for four years – they knew one another, had lived together, partied together, helped one another with field work and exam prep; they'd survived certain tough courses, not to mention tough profs, together. Their attitude was, Why not support one another, share the work! I'll give you this job, you give me that one.... We may be competitors, but more importantly we're in this together."

TOOLS OF THE TRADE

James Ferguson • 360-Degree Coverage

When James Ferguson entered the Erindale Survey Sciences program in 1979, the first GPS satellite had been circling the earth for barely a year, and few people outside of the professional physics and engineering community knew it existed. Ferguson, however, become fascinated with this bold new technology, and, in 1983, he and another student got their hands on some research range data collected by Texas Instruments from the four satellites that were now aloft. "This was stuff that nobody else had," said Ferguson in an interview in early 2016.

In short order the pair was at work writing a computer program that would use the data to calculate what Ferguson calls "the whole range of parameters" for global positioning. "We'd read books about this – one in particular, written by an American geodesist. It was kind of the bible of GPS in those days. We had very limited computer powers—were on the old vacuum machines, punching cards and stuff."

In 1984, Ferguson took time off from school, hooked up with a survey company, and embarked on an experimental GPS survey for a boundary retracement north of Wakefield, Quebec. "We rented the GPS receivers in the U.S.—V-1000s they were called, developed at MIT: great big things that cost half a million each and weighed three or four hundred pounds. We mounted them on the backs of cars and Jeeps."

In the field, Ferguson and company found cedar posts from 1850 that were still scribed, and positioned the receivers as close to these as possible. In those days, we positioned eight points, and it ended up costing us $2000 for each point. But we did get our survey tied together, and from there we were able to determine where the original boundaries were supposed to be. At the time, this was all pretty radical; we were really excited about it."

Phoning home from Latvia, circa 1992.

When Ferguson returned to Erindale, he teamed up with an electronics engineer, Paul Mrstik, who owned a small company called Geoserve, which was committed to the advancement of GPS. "We worked hard developing our expertise and reach," recalls Ferguson, "and within a few years had taken this revolutionary technology all around the world. We worked with it ourselves, taught it to governments, to surveyors, to engineers. Oh, it was hectic. Often we'd be up all night trying to get these rented receivers that we used functioning. Fortunately, we hooked up with a guy who'd written a program that enabled us to do all our calculations through a personal computer without having to go to the federal government to use their mainframe. It may not sound like much these days, but back then it really helped us make GPS useful and accessible to our clients. The company worked in Australia, South America, Europe; we did a stint in Russia at one point – did a lot of teaching and training in the U.S. and Canada. Basically taught people how to use the technology, how to analyze and use their data."

By Ferguson's modest assessment, he and Mrstik, the engineer and the surveyor, were a good team. "We could present the whole picture – understood that people needed useful tools and software, that it wasn't just about a cool piece of technology for collecting data."

By the early 1990s, the pair had begun to see the possibilities for using GPS in photogrammetry – fundamentally for operating cameras on helicopters. "At that point," says Ferguson, "we founded a company in South America; there were places down there where you couldn't get on the ground easily, so you needed that remote technology. And we worked with air photo companies in Canada and the U.S. We'd install modified cameras in their planes; we developed software around all of this. Whatever we had to do to solve a problem, we did it. We sometimes got to a point where we'd say, 'Okay, this is really cool.' But we didn't want to linger, so we'd move on to something else. As a consequence, there always seemed to be somebody coming behind us who would make all the money on our work. But that was okay; we were totally exhilarated by what we were doing, and were making a good living too."

By the end of the millenium, Ferguson and Mrstik had tied their ambitions to LiDAR, with the purchase of an early five-kilohertz Optech unit, with which they began both mapping and surveying. True to creative form, they eventually began developing their own LiDAR sensors. "We bought inertial systems and GPS from NovAtel and designed a system that would go on a helicopter. It was mainly for low-altitude high-accuracy power-line mapping and was one of the first commercially available systems. Some of them are still flying today, fifteen years later."

It wasn't long before Ferguson and Mrstik began to suspect that their helicopter-based LiDAR systems would perhaps work on the ground. "One day, in Ottawa, we did a crazy test – hung the LiDAR unit out of Paul's van, drove it around, and came home with this kind of point-cloud – no positioning, just a scan of all this stuff." But it would take a contract to map a highway in Afghanistan in 2003 to move them seriously into mobile LiDAR. "Originally, for that job, we were going to take our helicopter-borne system and mount it on an armoured air ship. Then we discovered that that would be extremely dangerous in the context.

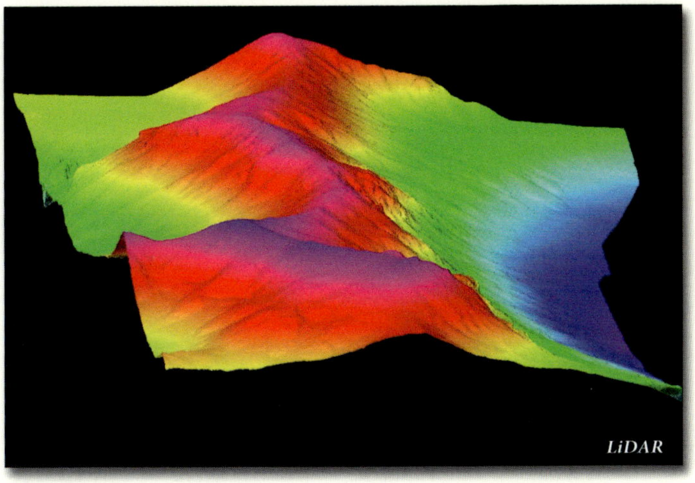
LiDAR

So we took our airborne system and, hearkening to our experiments in hanging the unit out the car window, put it on a pole on the back of a truck, and simply drove down a road as a test – and then flipped it over and drove back the other way. And it worked! So we developed the necessary software, got everything functional, and used the same method in Afghanistan – ended up with a system that we affectionately called Side Swipe.

"The problem was that when we got it into broader use, we had to go back and forth on a road three or four times, which wasn't economical. So we designed a system that had four lasers and gave 360-degree coverage. We sold one or two of those in 2006 and 2007 – probably the first mobile LiDAR scanner commercially available. Of course, since then, mobile LiDAR has taken off exponentially – much faster than GPS did in its infancy. Already we've got geiger LiDAR and flash LiDAR and photon LiDAR – have software that allows us all this amazing 3D point-cloud stuff. Funny thing is that most people still want nothing more from mobile LiDAR than a 2D map with elevation.

"Myself, I was never happy getting anything less than the maximum out of the resources and technology that were available. And yet I admit that I don't understand much of anything technical anymore. But I help run a company that does. And that's enough at this point. I've been surveying for 35-36 years. It showed me the world... and in a sense showed me myself. I'm not sure what more you can ask from a career."

Born in Cooksville, Ontario, James Ferguson was raised in Renfrew, by parents who immigrated to Canada from the U.K. By his own assessment, he "went into surveying so that he could be outside and his own boss." Today, James is general manager of Tulloch Mapping Solutions in Ottawa.

WDHS

Susan MacGregor (using using theodolite) and Elizabeth Campbell, at the Frost Centre, Dorset, circa 1986. Courtesy, AOLS archives.

CHAPTER SEVEN

The Age of Revolution

Unknown to most Ontario surveyors during the late 1960s and early seventies, change of another sort was quietly taking shape in the far southwestern part of the province. There, a surveyor named Bill Setterington had convinced his wife Lorraine that rather than merely assisting him on weekends, when she could take any or all of her three children into the field, she should take out articles, learn the profession, and become a surveyor. Until then, one hundred percent of the surveyors in Canada were men, had always been men. However, in May 1969, the percentage fell to 99.99, when the assertive and intelligent young woman from Leamington became the first woman in Canada to gain a surveyor's accreditation. During the years that followed, she also became a key, and sometimes controversial, figure in the larger history of Ontario surveying.

As we learned in an earlier chapter, Lorraine rather quickly found her way into the secretary's chair at the AOLS offices – and, shortly thereafter, as a result of her commitment, into the role of executive director of the AOLS. Which, according to a long-time associate, "is when things got interesting – both for Lorraine and for the Association."

In the words of the Sudbury surveyor, Anne Cole, "Lorraine was all about standards, about holding the profession accountable: you *will* behave in a professional manner; surveying *will* be correct; it *will* be well-researched, you *will* serve your clients; this is *not* some self-serving club in which you make up the rules as you go along. And if you're not doing that, you're out of here! It was that insistence on accountability that defined her term as director."

During her first dozen years in the role, Lorraine prepared more than thirty disciplinary cases against land surveyors, each involving what was perceived to be incompetence or unethical behavior.

In May 2015, she said, "I believe we removed eighteen or twenty licences in the time I was director. It was tough be-

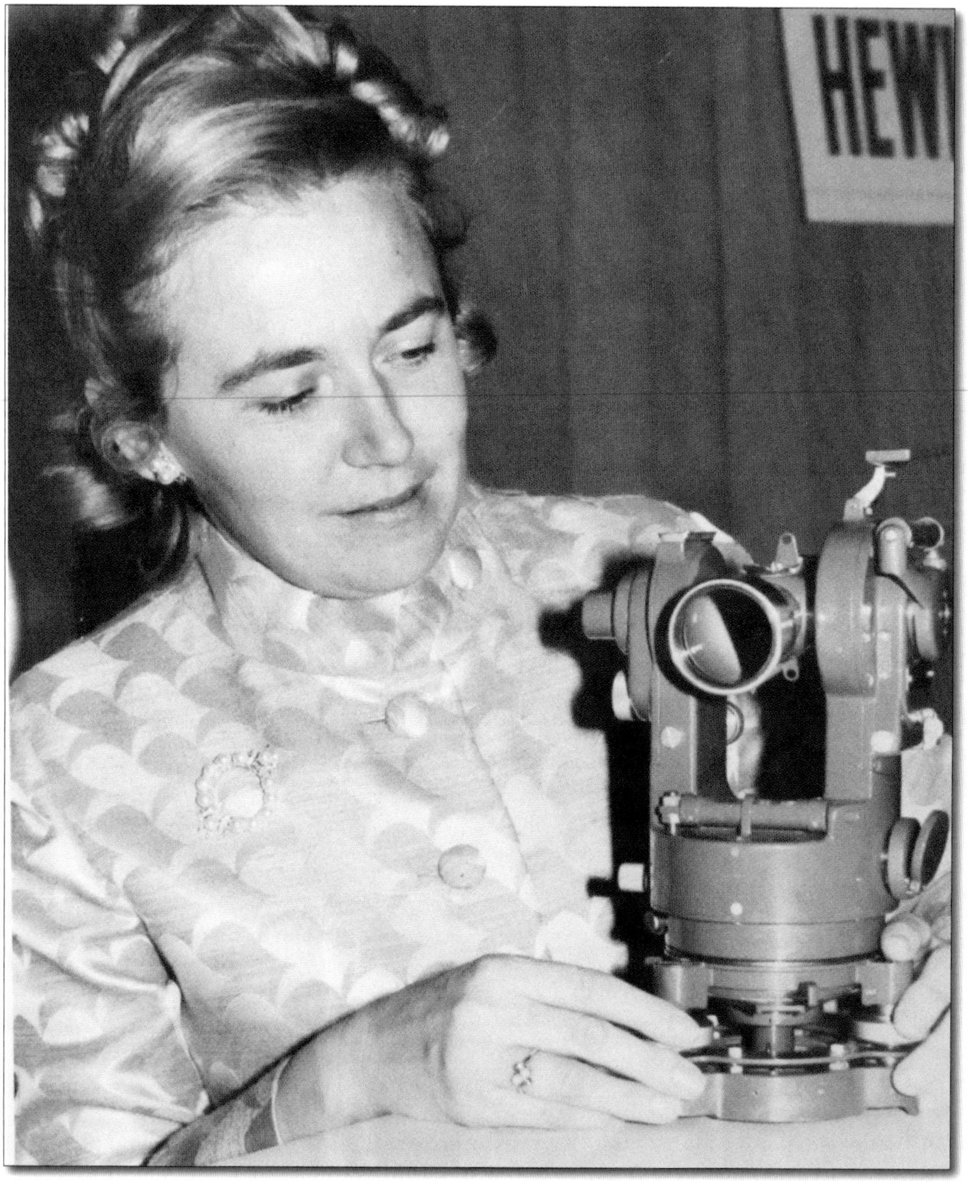

cause I knew the guys, and they all knew me. Once, it was a fellow who gave a toast to my husband at our wedding. But we had to do it. The Surveyors Act stipulates that the Association exists to protect the public, not to protect the surveyor."

Along the way, Lorraine managed to outlaw practices such as the notorious "drive-by" survey, by which a surveyor, hired by a developer or lot owner, would sign his name to a survey plan after just a cursory glance at the site it represented, or with no glance at all. And against a tide of resistance, she worked with the AOLS council to establish a review process that would subject Ontario surveyors to periodic assessments of their work.

"Lorraine's changes may have seemed harsh to some," said a past president of the organization. "However, in some cases she also made things fairer for surveyors who had messed up. For example, she broadened the disciplinary process so that rather than looking at just one bad survey that had provoked a complaint, the discipline committee examined surveys selected from five years of a surveyor's work – which of course gave a more indicative and fairer picture."

Asked recently if being a woman made reforms more difficult in a predominantly male industry, Lorraine responded, "No! It made it easier! I was an executive mother sort of thing. I'd get a phone call at 11 o'clock at night saying, 'Can you talk to me – I've got a real problem?' I don't think they would have done that with a man."

As the AOLS historian John Ladell wrote in 1992, "With her eye on the

Opposite page: Newly commissioned surveyor, Lorraine Setterington, OLS #1234, examining the latest in theodolites during the early 1970s. Lorraine arrived in the Association's executive director's office in 1976 with improvements in mind, and did not waver in implementing them. Right: Laurentian University in Sudbury capped Lorraine's extraordinary career in June 2009, when it made her an Honourary Doctor of Laws. In introducing Lorraine at the ceremony, Sudbury surveyor Anne Cole (centre), noted her "intelligence, energy and integrity," and described her as "tireless in promoting education and professionalism for land surveyors across Canada, the United States and internationally." Laurentian President, Dominic Giroux (right), bestowed the degree.

future, Petzold [Lorraine was by this time married to the surveyor "Red" Petzold] became heavily involved with high schools, community colleges, and universities in an ongoing effort to attract and retain high-quality recruits to the profession." In the interest of standards and the image of the industry, she lectured to surveying associations in Canada and the U.S., and occasionally farther afield. Back home, with Council's directive, she introduced mandatory liability insurance for surveyors, and at AOLS headquarters built up the Association library. "It helped that I was a workaholic," she said.

"It also helped that she was very smart and a good writer," said Jack Young, who was president of the AOLS in 1988. "I'd be asked to go and speak at one event or another, sometimes at the last moment, meaning that I didn't always have time to properly prepare. Lorraine would say, 'Don't worry about it; I'll pull something together for you. And she would – in no time. And the speech would hit the bullseye."

Throughout her time as executive director, Lorraine participated extensively in Canadian Bar Association and Law Society seminars, and gave countless lectures to lawyers' groups, all aimed at improving the relationship between surveyors and the legal profession, as well as encouraging proper application of boundary law.

During one of her addresses to a county bar association during the late 1980s, a young lawyer in the audience challenged her repeatedly from the floor, eventually demanding to know what her qualifications were to be instructing lawyers and why anybody should take her word about anything. "He'd been rude and arrogant all evening," said Lorraine, "and at that point I told him that I believed I was qualified to deal with the likes of him because I used to teach kindergarten. He stormed out, and I got a round of applause from everybody else."

In 1989, armed with the knowledge and perspective that had gone into her lectures, Lorraine served on the steering committee to create Canada's first textbook in survey law. Meanwhile, in 1987, she had been presented with the Champlain Award "for outstanding service to the land surveying profession," and, upon her retirement in 1992, was given the Association's "Professional Recognition Award." In June, 2009, Laurentian University conferred on her an Honourary Doctorate of Laws.

"I still hear complaints about Lorraine," says Jack Young, "about how she went too far in cleaning things up, introducing new standards. As president, I had some disagreements with her myself. However, I think for perspective, it's wise

Maureen Mountjoy, editor of the Ontario Professional Surveyor *(see opposite page), happily acknowledges the help of her husband, Bob Mountjoy, OLS, and of Bill Buck, the AOLS registrar, in proof-reading the galleys of the Association's "quarterly miracle," the Ontario Professional Surveyor.*

to remember that during the twenty-odd years since she retired, the standards she introduced have not been relaxed by a succession of executive directors and councils. In fact, they've been tightened, raised. So, it's impertinent to say she went too far."

IN THE SPIRIT OF HER 16 YEARS as executive director, Lorraine's last major responsibility on the job required her to hasten to Ottawa in 1989 with the Association's lawyer, Brian Campbell, in order to fight a charge that the surveyors of Ontario were in collusion on their fees. If true, they were in violation of the Competition Act, and were subject to a fine that, according to the Act, could have been set as high as a million dollars.

"It would have ruined us!" said Lorraine. "When we got to this meeting, a guy came in with a stack of files and plunked them on the table and said, 'We're filing criminal charges against all of you.' By this time, our president, Michael O'Sullivan, had joined us; he worked in Ottawa. I don't know how Brian and I drove home; it was such a shock. The Competition Bureau was waging war against price fixing in various professions, and in us they'd picked the smallest profession, figuring they could make an example of us and that we wouldn't contest it."

Immediately, Lorraine called a meeting of the Association's seven hundred members, the purpose being to discover if they were willing to put up the money to fight the charges. "More than six hundred turned up," said Lorraine, "and just one person voted against supporting a fight. Afterwards, I asked him why, and it turned out he didn't have a clue what was going on.

"So we fought it – hard. What the government wanted, besides this massive fine, was for us to admit our wrongdoing and apologize on the front page of the *Globe and Mail*. But none of that happened. In the end, we came to an agreement with them that they would review us once a year. It was a soft compromise on their part, but there wasn't really much else they could do: we had *never* said to our people, 'You have to charge X amount for a survey.' Mind you, we *always* encouraged our surveyors to do a thorough job and to charge for the time it took to do things right. But, oh, it was a gut-wrenching period; it went on for a couple of years. When the final agreement came down and was signed, I was gone; the fight was over. New people were in charge."

The Quarterly Miracle

The New York Times *has been called "the daily miracle," in that somehow, against the odds, the thick newspaper appears every 24 hours, without fail. Likewise, the* Ontario Professional Surveyor *might be called "the quarterly miracle," based on its own faithful and "miraculous" occurence every three months. The difference of course is that the Times has an editorial and production staff of thousands, while the OPS is managed, edited and marketed by a single soul, the dedicated Maureen Mountjoy, who is able to commit a mere quarter of her working time to creating the glossy and colourful magazine. "I do have the help of a wonderful production manager, Lori Veljkovic," explains Maureen. "And I'm so grateful always to those who write for us. But basically I myself assign the articles, bug the writers to get them in on time, edit them, find appropriate photos and illustrations, write whatever other parts of the magazine need writing, create the cover, sell the ads, proofread the stories and galleys..." She pauses for a laugh. "By that time I'm busy wondering how on earth we'll ever get the next one done."*

The publication began in the 1950s as a newsletter, and for a decade or more during the 1970s and 1980s was edited by the well-remembered writer and humourist, Andy Gibson, OLS. "During the years after it became a glossy magazine, it consisted mostly of reprinted stories from other publications," says Maureen. "My idea when I took it over in 2001 was that we have so many talented surveyors doing fascinating projects, we should be showcasing them, plus doing educational stories that surveyors could apply to their work."

Asked about her biggest challenge in running the publication, Maureen responds, "Getting the writers to say something interesting. The whole thing is to speak up, to stimulate our surveyors, make them feel positive and progressive in their work."

Prompted to name her favourites among the 60-plus issues she has edited during 16 years on the job, Maureen cites without hesitation the Fall 2007 issue (whose cover commemorated David Thompson's work in Muskoka in 1827); the Fall 2005 issue (commemorating 125 years of surveying in Manitoba), and the Spring 2012 issue (whose cover shows Sergeant-at-Arms, Julia Meldrum Smith, OLS, impersonating Queen Juliana of the Netherlands at the 2012 AGM in Ottawa) – all depicted above.

The Cambrian Society, a group of early graduates of the Erindale Survey Science program, has existed for nearly 40 years for one reason only: to make its annual cross-country ski excursion into the wooded uplands of central Ontario. Members are seen here at Black River during the late 1970s. From left: Don Austin, John Clark, Alex Osborne, Jim Zucchiatti, Rob Harris, Peter Bull, Blain Martin, Tim Hartley, Tom Packowski.

Despite its fame and efficacy in the grand scheme, the Erindale program too would be gone by the end of the millennium, having lasted just 26 years. There is still disagreement over exactly why it faded in the mid-1990s and was terminated in 1998. One long-time supporter of the Erindale program believes it was let down by the upper administration of the AOLS: "They disbanded Fred Pearce's education committee, and in effect pushed Fred aside. It was that committee that had done the most to found and inspire what was going on at Erindale. With its small number of students the program wasn't a money-maker – I don't think it ever had more than about a hundred students at one time, including all years; and because of that it needed nurturing. Fred and his group gave it that. When the committee was disbanded, there was lip service to supporting all survey programs in the province, but in effect the break-up of the committee was an abandonment of the special connection with Erindale."

Current AOLS Executive Director, Blain Martin, says, "I was on that committee, and I don't think Fred Pearce was in any way 'pushed aside.' He was getting older and was glad to relinquish responsibility, even though he was disturbed to see the program die."

Gordon Gracie believes the shuttering had far more to do with university politics and an aging faculty, several of whose key members retired during the mid-1990s. "Bob Gunn and Dave Lambden both retired, then I retired," said Gracie in May 2015. "By that time, our hydrography expert, a chap named Kapoor, had died – as had our geodetics professor, Attallah Wassef."

Anne Cole • Renegades and Outliers

"**Before I started in Survey Science at Erindale,** I thought I should have a summer of experience surveying, in case it turned out to be something I didn't want to do. So, that April, 1976, I started working in the field for a guy named Murray Maher, a bit of a renegade in the Sudbury surveying community. Back in the winter, I'd written very proper letters to all the surveying firms in Sudbury – there were four at the time. Murray responded, which is why I had the job, but one firm never did, and two of them wrote me in March saying, "We have no requirement for field staff this summer." Which I later learned was complete B.S., because nobody knows in March whether they'll need extra staff. It was basically a kiss-off – we're not hiring a woman. But I didn't know that then.

"The quintessential question was whether they were just saying they didn't want women around, or whether they'd thought this through and truly didn't believe women could slog through the bush and swamp.

"I haven't explored this academically, but I believe there was a fear factor, and maybe there still is – basically that sexuality can disrupt the equilibrium out there in the survey camp, or in the field generally. And of course one of the ways of controlling it is simply to have no women around, or at very least segregate the sexes. To my mind, looking back, that fear was a far bigger barrier to women entering surveying than any belief that women couldn't actually do the work.

"What complicated things was that I came into surveying when it was more physically demanding than it is today. I'm not saying it's not physical now, but the equipment today is so much different. When I started we were carrying a lot of heavy items way back in the bush. And there were other little complications – even things like going to the bathroom with a bunch of guys around. Or you have your period. These things can be limiting.

"On the other hand, one of the things I've always said about women entering surveying in the north is that northerners are renegades and outliers to begin with, so the men were probably better disposed to change than they might have been in some parts. Meanwhile, I'd say that 95% of the men I ever worked with are thoughtful, honourable people. Which is pretty much indicative of the population anywhere. Maybe it's my bias, but I'd also say there's a higher proportion of good men in surveying, generally, than in the broader population. They're really a very decent lot, well connected to the earth – just solid people."

In 2015, Anne served as president of the Association of Canada Lands Surveyors. She lives and works in Sudbury, where she is a a senior advisor to Tulloch Geomatics Inc. and a consultant to the industry through her company, J. Anne Cole Surveying Ltd.

Eric Ansell • One Canoe, Two Pontiac Hoods, Nine Drops of Rain

"**Back in the nineties, when I was in private practice,** we were doing a retracement survey north of Sudbury – running the line between two townships, about thirteen miles through the bush. And of course up north you could only drive so close to the line, then you had to walk. So at seven o'clock every morning we'd get out of the truck and start tramping through the bush until we hit the cut line that we were running. Then we'd walk up the line for several miles until we got to where we had left the equipment the night before, and we'd continue cutting and running the line, looking for evidence of the old line, and so on. In all, it was a pretty grueling process – lots of long days.

"Anyway, one day as we were in the midst of this, we came upon a good fiberglass canoe that was just sitting in the bush beside a fairly substantial creek, with no sign of anybody around or having been there. It was actually kind of mysterious. But being a surveyor and at least somewhat resourceful, it occurred to me that it might be a good idea to use the thing to paddle our equipment up the creek, which just happened to run pretty much alongside our line.

"So we did! Used this canoe steadily for a couple of weeks – saved ourselves a ton of work, imagining all the while that nobody was any the wiser.

"But these northerners are a pretty smart bunch – and also very accommodating and helpful, especially in the bush. And eventually one night I got a call at the motel from a guy who asked me very calmly if I was the surveyor who'd been using his canoe up and down the creek for the past couple of weeks. He hadn't missed a moment – knew exactly what we'd been up to with his canoe all along. When I told him that, yes, indeed, I was the guy and started into the apologies and so on, he said 'Oh, that's okay – you obviously needed it. But could I ask you one thing? Would please make sure to have it back where you found it by Friday afternoon? I put it there in preparation for a hunting trip, and I leave on the weekend.' And he thanked me in advance for my kindness. And I thanked him for his. The only difficulty for me was that the trip into the woods that Friday was a little harder than normal, because I was carrying a good big case of beer that I left in the canoe when we put it back in its place.

"During another retracement survey, this same crew and I came upon a fairly large beaver pond. So I sent two guys around to the east and I went to the west to figure out which way was the best to get around to the other side. Just as I began picking my way through the brush, I spotted what at first looked to be a boat on the shore. I thought, Great, exactly what we need! I went to investigate and started to laugh, because in fact it was a sort of sleigh made of two old 1955 Pontiac hoods welded together. It may have been some sort of vessel but had probably been used for hauling stones or gravel or something overland. I was about to walk away and I thought, Wait a second – I'm a surveyor; I can make something out of this. And I eased this rusty old thing into the water. And *voila*! It floated! Meanwhile, the other guys had come over, and we very gingerly stepped into our makeshift canoe, got our centres of gravity as low as we could get them, and ended up poling across the beaver pond – as daring as the surveyors of old.

"On yet another occasion, we were out doing a private survey and there happened to be an MTO survey crew doing some highway work right there adjacent to us. Mid-morning, it started to rain, and the MTO guys jump into their vehicle right away. We continued working, and a while later, during a break, I went over and said to them, 'At what point do you decide that it's raining hard enough to quit?' The crew chief said, 'We draw a one-inch square on the windshield and if there's nine drops within that one-inch square, it's raining enough that we have to cease operations until it lets up.' I said, 'And how long do you wait for those nine drops to accumulate in the square?' – to which he said, 'Oh, sometimes we have to sit here all day.'"

Eric Ansell graduated from the Geodetic Science program at Ryerson University in 1976 and was commissioned as an OLS in 1979. From that point, he was in private practice for three decades, served as president of the AOLS in 2013, and was the coordinator of crown land surveys in the office of the Surveyor General of Ontario from 2010 to 2016.

As the Erindale program blossomed and then collapsed, highways continued to get built, farms became suburbs, AOLS members continued to travel the globe. Alec McEwen is seen here during a posting with the United Nations in the Seychelles, where he conducted triangulation control surveys and trained local technicians in the science of basic surveying.

Throughout the early 1990s, the surveyor and lawyer Izaak de Rijcke had been with the program as a lecturer in survey law. "Toward the end, we tried to bring him on fulltime," Gracie said. "But understandably he wanted tenure if he was going to leave his practice, and the president of the university, another lawyer, wouldn't give it to him because, at the time, Izaak didn't have an advanced degree. That was a big blow to us. Izaak was a very good teacher, with a broad range of curiosities and interests."

The program did hire a couple of new professors in an attempt to retrench. "But they were academics, not surveyors," Gracie explained, "and that hurt us. By the mid-late 1990s, the only surveyor left was Jack Young, who was of course excellent, did a great job for the program for many years, but he couldn't do it on his own."

Gracie believes the college would have been happy to preserve the program. "Unfortunately," he said, "there was such an administrative void at the time of all the retirements, the college felt the program should be under the wing of a larger, existing discipline. So they transferred it to the geography department. The problem was, the geographers weren't surveyors. They were committed people, but in the end surveying didn't mean anything to them."

Gracie points out that the 1990s also happened to be the Mike Harris era, a time of pinched education budgets and hence little money to hire new people. "Surveying is a small profession compared to law and medicine and engineering," concluded Gracie, "and in the end we just weren't able to muscle our way through and keep it going."

GIVEN THE PROMINENCE of the Erindale program, it might easily be overlooked that between 1972 and 1982 the engineering department at Ryerson University ran a parallel program that graduated perhaps a hundred surveyors, compared to the 450, total, that came out of Erindale. "The Ryerson grads were fine surveyors and ultimately excellent members of the AOLS," says Blain Martin.

The 1976 class of graduates from Ryerson University's survey program (1972-1982) produced four AOLS presidents. No class at Erindale can make the same claim. Those presidents, then and now, from left: Doug Culham (2006), Dave Urso (1999), Eric Ansell (2013) and Walter Kowalenko (2010).

John Halsall, a member of the first graduating class at Erindale, at work in the Porcupine River Valley in Northern Alaska, 1976.

Surveyor Eric Ansell, who served as president of the AOLS in 2013, was a graduate of the first class at Ryerson. "Four Association presidents, including myself, came out of that first group of sixteen grads," Eric said in March 2016. "I'm not sure the first Erindale class, which was quite a bit bigger, can make that boast. I think we probably resented the Erindale program ever so slightly, because it got so much of the attention – it was definitely the darling of the AOLS."

During the years since, Ryerson's engineering department has run a progressive and energetic geomatics program and continues to send graduates into surveying. Likewise, York University's Lassonde School of Engineering has evolved an active and respected program in geomatics, many of whose graduates have gone on to become commissioned land surveyors.

Apart from advances in education and standards, the Erindale years were an era of hard-won coalescence for surveyors in Ontario – a gradual amalgamation of cadastral surveying with photogrammetry, geodesy, hydrography and geographic information systems (which is to say the use of amassed data in everything from defence systems, forensics and ecological assessments to the measurement, say, of crop yields, oil and gas reserves, and the slow, grinding migration of the earth's tectonic plates).

For Ontario surveyors, it was also an era of adventure, and in some cases idealism. During the mid-1960s, for example, the Sarnia surveyor, Jack Monteith, plied his talents for several months on the Indian Ocean along the northwest coast of Australia, where he and a partner had contracted to lay out navigation markers for a new industrial harbour. "One day we flipped our boat on a reef and had to wade several hundred metres to shore, surrounded by sharks," he said in June 2016. "It helped me appreciate black flies."

Likewise, in 1981, John Halsall fetched up in Egypt for 18 months, in charge of one of the largest hydrographic surveys of its time, on the Nile River. John's hydrographic work eventually took him to the corners of the planet, including a mid-1990s stint in the South

Pat Hills • Wonderland

"I remember being on the Port Weller Drydocks in St. Catharines when they brought an ocean liner in, with a big hole punched in it. The mission was to patch it. But the company I was working for at the time couldn't construct the patch based on an original drawing of the vessel, because once you put a ship 700 feet long into the water, and then bring it into drydock, its shape has changed. So we had to do an 'as-built' – in effect, a 3D model that shows something as it actually exists, not as it was configured at the drawing board. Our people worked up this model of the whole compromised area, and built the patch based on the model. That way, the patch was perfectly contoured to fit the hole, and when they built the heavy framing to get it all up there into place, they didn't have to take down again."

During the 1980s, Pat Hills worked as a surveying and engineering technologist. But the recession of the early 1990s pushed him toward the business side of his field, and these days he is in management with the Canadian firm Cansel, providing equipment and what he calls "measurement solutions" to surveyors and engineers. "One of the most exciting things about the job," he said in an interview in late 2015, "is that I never know what a new day is going to bring. I'll get a call from a surveying firm – they'll say, "Pat, the province has asked us to figure out if the new overpass at the 407 is moving; chunks of concrete are falling off. How are we going to sort out what's going on?

"Or a company will be monitoring the removal of a townhouse from a row of them, maybe a hundred and twenty years old. Once the structure is gone, there is a chance the adjacent buildings will move. What they want to know is how best to measure this movement to ensure stability and safety.

"Or I'll get a call from a surveying or engineering firm – they're profiling a rock cut up north. They don't want their people out on the road, and of course you can't always just walk up to an eighty-foot-high rock cut. Plus, they need two-millimetre accuracy. So my job becomes figuring out a way to help them get this done."

It is work that frequently requires both intuition and a detailed process of discovery. "A mining company might need to measure volumes of ore removed from underground," says Pat, "or make sure they know where their tunnels are going. In many cases these days, you can tell what's underground from the surface using ground-penetrating radar. But you don't usually reach a solution without a fair bit of discussion with your client, and by trying things out.

"In other cases, a drone might be the answer to a client's requirements. They might have huge stockpiles of ore or mine tailings that need regular measurement. If the material is unstable, it's unsafe to send somebody up on it. But with a drone, you can be a couple of kilometres away, safe in a field of sunflowers."

Pat allows that he's had some fun at his job over the years. For example, he was approached once by the president of the Canadian Water Ski Association, which was hosting the World Water Ski

Championships back in the 1990s. "He needed to position the gates on the slalom course within a certain tolerance, and anchor them, and also position the jumping ramp. He realized he needed somebody with an awareness of very precise measurement. I was just a rookie, but I went to work on it. Really enjoyed it. But I'll tell you, I was extremely nervous watching the championships – was afraid somebody on the slalom course was going to hit the side of the jumping ramp because maybe I'd positioned it the wrong way."

Pat has had a thrill or two as well at Canada's Wonderland, where he has provided technical equipment to the land surveyor who monitors the assembly and testing of the park's new rollercoasters. "When they were putting up the Leviathan and Behemoth," says Pat, "The surveyor was there right through the construction, riding the rollercoasters with the dummies, measuring everything, making sure the pillars and tracks were all in perfect trim. This is of course all to very tight tolerances. It's the kind of thing you call a surveyor for, because of his or her commitment to getting a job like that done right and safely; really no one else should, or even could, take it on."

Likewise, explains Pat, a surveyor is crucial to jobs in, say, the big nuclear plants, where safety and precision are both necessary and and strictly enforced. "Let's say they need new ducts from one end of the building to the other, and they have to run them through all the overhead plumbing and venting and electrical and mechanical equipment. What we'll do is scan everything and create a 3-D model, so that an engineer can design a fixture that can be installed without hitting anything."

Pat is enthusiastic in heralding progress in surveying and engineering in recent years – and the resulting expansion of humanity's awareness of the planet. "For example, there's no longer much of the planet that hasn't been mapped to a very high degree of accuracy," he says. "It was exciting last year when they discovered the ship from the Franklin Expedition under the Arctic ice. It happened largely because they were up there mapping the bottom of the ocean, the shorelines, and so on. Surveying is a business and a profession, but it's also a kind of adventure, during which one exciting discovery so often leads to another."

Bob Halliday calls his surveying adventure in Nunavut in 2005 one of the highlights of his 40-year career in the industry. "Given the right opportunity," he said in November 2016, "I'd move permanently to the Northwest Territories or Nunavut."

Above: Bob's photo of a barren ground caribou in its Arctic terrain.

China Sea, where for two years he worked on the construction of a pair of underwater pipelines, one of which ran 700 kilometres from the southern Chinese mainland to Hong Kong.

Throughout the early years of the new millennium, the surveyor and GIS expert, Zul Jiwani of North York, spent a substantial portion of each year travelling as a volunteer to underdeveloped parts of the world, applying his skills to the improvement of geophysical conditions that had led to a variety of natural disasters. In 2005, under the banner of the humanitarian aid group FOCUS, Zul spent three months in the mountains of central Asia (a zone encompassing parts of China, India, Pakistan and Afghanistan), where he advised on the installation of an early warning system to help villagers cope with the landslides that for centuries had wiped out one village after another.

During the same period, the surveyor James Ferguson, widely recognized for

One of the glories of surveying has always been its potential for memorable experience: a sighting of bear cubs in the boreal forest; the mysteries of the catacombs beneath the city or the mountain peak as explored by the UAV... and everything of interest in between.

his work in digital technology, travelled to various parts of Asia, eastern Europe, South America and elsewhere, teaching early applications of GPS and LiDAR, developed by Ferguson and his partner, Paul Mrstik.

Exotic projects are not of course exclusive to distant continents. Closer to home, Paul Gregoire (working for Marshall Macklin Monaghan) conducted boundary surveys during the mid-1980s in the splendid Arctic wilderness that a dozen years later would become the Territory of Nunavut. In 2005, the Espanola surveyor Bob Halliday travelled to Nunavut to place permanent markers on mineral claims that had been staked along the Arctic Circle. Faithful to his rural inclinations, Bob's written reminiscences about the venture devote less attention to actual surveying than to Arctic vegetation, sunsets and wildlife – to species such as caribou, wolves and muskox, as well as to Arctic loons, with whom Bob reported enjoying daily conversations. "I made the mistake of telling the guys in the gang that the loons always said hello to me," he reported in early 2016. "From then on they asked me every evening what birds I'd been talking to *that* day."

THE WANING YEARS of the 20th century witnessed a range of geodetic and data-focused projects in Ontario. One of the most prominent of these was a Ministry of Natural Resources initiative called Ontario Base Mapping, which coordinated a broad range of cadastral data into a single digital resource base. From that base, detailed coloured maps of any part of Ontario could be printed on demand, at any scale. "I'm still amazed at how quickly all this happened and then was all overtaken by, say, Google maps and GPS, and so on," noted Eric Ansell of the Surveyor General's office, in mid-2016.

From the perspective of the AOLS, the period fell largely under the administrative terms of executive directors Carl Rooth, Murray LeGris and Jim Statham, all of whom kept a rather low administrative profile relative to those of their predecessor or successor (whose lively directorship we are about to touch upon.) "They were quietly conservative guys," says a Council member of that era. "The Association had experienced some fairly radical redirection. The nineties were a time for just staying on course."

As previous parts of our book have made clear, the new millennium brought a new order of change – an ascendency

The vessel depicted above is engaged in the wide-swath mapping of the floor of the Great Lakes. What was once done with a plumb line over a period of years is now done remotely, in a matter of hours, in 2000 times the detail, from the surface.

of GPS units, satellites, robots, target-seeking UAVs, self-actuating total stations, mobile LiDAR devices, ultra-sensitive cameras, underwater mapping units, all of which are functionally inseparable from the advanced computerization on which the industry is now dependent.

However, if you were to ask Blain Martin about the biggest advances since he assumed his role as executive director of the AOLS in 2009, he would perhaps apply the question less to the actualities of surveying than to the functioning and concerns of the Association. "The story here over the past few years," he said in March, 2016, "is the *huge* improvement in communications among members. There are just so many opportunities these days for them to come together, at least electronically." Among other virtual meeting places, Martin cited the Association's monthly "webinars," its biweekly electronic newsletter (a brisk and entertaining missive), and its impressive interactive website. "Thirty years ago," he said, "the executive director had four secretaries, who spent hours every day typing up the stacks of letters she sent to members. You sent one out; a week later you got a reply – if you were lucky. Now we fire out a hundred emails a day – sometimes five hundred. And the responses are instantaneous. We've got surveyors engaged in the Association right now who five years ago had never attended a regional meeting or an AGM, and may have known three other surveyors in the province."

One can accept Martin's assessment that the transformation is the result of rapidly advancing communications technology. But it is also a function of Martin's management style – a relentless, brotherly, talkatively supportive rain dance: a way of operating that, combined with his high expectations for everybody within hailing distance, keeps the Association's executive sharp, its committees motivated, and its membership informed and involved.

"Blain is a *very* resourceful dude, with a great sense of opportunity," says a long-time associate. "He listens to everybody, then does what he wants – and most often does it right."

A member of the first graduating class

From left: Executive directors, Carl Rooth (1993-99), Murray LeGris (2000-05) and Jim Statham (2006-09). This trio of land surveyors guided the Association from its centennial year in 1992 until 2009, when Statham's Erindale classmate, Blain Martin, took on the executive director's role.

at Erindale, in 1976, Martin articled with one of the province's oldest firms, Speight, van Nostrand & Gibson. For 17 years, beginning in 1983, he was owner and president of Blain Martin Surveying – a period that culminated with his earning an MBA from Queen's University. Before being hired to lead the Association, Martin worked for seven years in management for NAVTEQ, a Chicago-based company that creates electronic navigable maps of the sort found in a variety of GPS units.

Asked about his vision for the AOLS, the director says, "Everything I've tried to do since I got here has in some way been aimed at bringing recognition and respect to the profession. Surveying is core work in this culture, and so few people know anything about it."

Martin has also committed his directorship to what he calls "the business of the Association" – to balancing the books, attracting revenue, spending money not just to create programming but to create more money. "It's so important to have some resources around to develop your programs and to fund a special project once in a while," says past president, Jack Young.

More than any other factor, the director attributes the success of the Association during recent years to what he calls "a fantastic staff" – several of whom, it must be said, were hand-picked by Martin.

Whether by selection or coincidence, the current coterie of administrative personnel includes a poet, a professional concert pianist, an artist/designer, a former television personality, and a Russian linguist currently completing an MBA at the Schulich School of Business at York University. They have come to Canada from Ukraine, the Philippines, Siberia, Lebanon and Brazil. "Two or three were even born in some exotic place called Canada," jokes Martin who, at the time he hired or met his creative charges, was unaware of their artistic backgrounds.

Meanwhile, they show him their loyalty and respect, and he returns it, adding openness and approachability along the way. "Everybody speaks up when they have something to say," submits one staff member. The atmosphere around the office is sufficiently informal that the Association's member services coordinator, Penny Anderson, frequently brings her dog Barkley to work, meaning that visitors to the second floor are at times greeted by a restrained woof or two upon their arrival. Which is all fine with Martin, who occasionally brings his own dog, Trinity, to work and who cites research indicating that the presence of a dog in an office reduces stress and creates a sense of goodwill among employees.

Martin is flanked professionally at headquarters by a quartet of talented land surveyors that includes his one-time classmate at Erindale, Maureen Mountjoy, who has been a recurring and energetic feature of our story, and who wears half a dozen AOLS hats – among them those of assistant to the registrar, house

Mark Tulloch • Power

Mark Tulloch, the 35-year-old president of Tulloch Geomatics Inc., has been called a visionary in the field of contemporary surveying. The family firm was started by Mark's dad, Mike Tulloch, in Thessalon in 1991. But the late 20th century was not an easy time for a surveyor in a northern town, where work was often difficult to come by. "The recession of the early nineties pushed the original company pretty close to the brink," Mark said in May, 2015. "There were times when my dad considered closing the office and simply running things out of the garage at our house."

But with a change of business philosophy and the arrival of Mark full-time with the company in 2006, the firm regained its footing.

Today, it is an expansive and profitable operation with a splendid head office in Huntsville, as well as ten offices in other parts of the province, and more than 275 employees. The firm has surged not just in size but in its application of innovative technology, in business and professional practices, and in recruitment and education.

"We learned two very important lessons from the nineties recession," said Mark. "The first was that doing cadastral surveying and nothing else can be very limiting when times get tough. So, having scraped through the economic downturn, we quickly diversified into other areas. Now, if there's a fall-off in, say, land development, where most cadastral surveyors practice, we can refocus our efforts in the public sector, and vice versa, and are better able to ride out the waves.

"The second lesson is that you can't get too confident in your success." It is a lesson, Mark allows, that the company is constantly re-

learning and reapplying. "We'd say we're a cautiously entrepreneurial company," he explains. "Admittedly, we're always looking for opportunities to grow. But they have to be the right opportunities – financially manageable. As a company, our strong preference is not to borrow money – for anything. We pay cash, whether it's for trucks, GPS equipment, or sometimes whole survey firms. That way, if the business or any part of it fails, the worst that can happen is we close up shop and life carries on."

Tullochs is perhaps best known for its engineering surveys and project administration, particularly in the field of transportation. But the firm has also emerged as a specialist in renewable power facilities. With that in mind, Mark was asked recently to provide a glimpse of the company's operations by describing its work in the installation of Ontario wind turbines. It is work that began for Tullochs in 2004, when plans were completed for the Prince Wind Farm, one of Ontario's first such facilities and at the time the biggest wind farm in Canada, just north of Sault Ste. Marie, on Lake Superior.

"Back then," said Mark, "Canadian contractors were totally unfamiliar with wind turbines." As a result, Brookfield Power, the farm's owner, hired two American firms for construction and engineering. "The problem was that these guys were accustomed to putting up turbines on farmers' fields – not in the wilderness north of Superior."

For that reason, on a day in early 2004, Tullochs got a call from the U.S. contractor, who needed local support designing roads over rock outcrops and through swamps and spruce forests. "Back then, we had no idea what a wind farm even looked like," said Mark. "But we thought the idea was pretty exciting, so we partnered with Mortenson Construction, and for the next three years, 2004 to 2007, provided the project with a whole range of surveys and civil engineering design work."

For convenience, when the job began, Tullochs opened a small office in the Soo, then bought out surveyor Bill Bolan, establishing a branch that is now one of the company's largest. "It houses our engineering design group, our civil engineering group, and a couple of survey crews – about thirty people in all," says Mark.

By the time construction was finished in 2007, Mark and others at the firm had, in his words, "become passionate" about the wind and solar industries. "We still had no idea what the Ontario market would offer in wind but we'd nevertheless become active members of CANWEA, the Canadian Wind and Energy Association. In 2006 when we went to our first smallish CANWEA event, wind power was still this kind of this hippie concept; hardly anyone in the country was investing in it. Then in 2009, the Ontario government tendered six more wind projects to further test the concept, and we were fortunate to get one of them."

When in 2010 the Liberal government initiated the Feed-in Tariff or FIT program, 147 renewable energy projects were launched. "All of a sudden," said Mark, "everybody and his brother was involved in the wind and solar industry. During the course of four years, we saw the CANWEA events grow from 400 to 2500 people. In the meantime, we'd formed a lot of good relationships with wind developers and contractors, so that we were able to capture a lot of the new work in various parts of the province. To date I think we've commissioned about a thousand megawatts worth of wind power in Ontario, and have commissioned another 650 megawatts of solar power, which is massive. At the moment, our market share of renewables in Ontario is around thirty-five percent."

Asked about the specifics of the company's on-site work, Mark said, "We design the roads to get to the locale; we do all the site design for the turbine locations – the ground beneath these turbines has to be absolutely flat. We do all the topographic surveys and the cadastral surveys, as well as the environmental investigations. For example, we count the number of birds and bats and reptiles on the site, all the rare species, both ter-

restrial and aquatic; we make sure we're not putting a road through a sensitive habitat, or dislodging an endangered species. Plus, we do a lot of geotechnical work – for example, at every proposed turbine location we drill down anywhere up to 30 meters to see what the soil and rock conditions are like. Geo-tech is a big application for us where renewables are concerned. We also get involved with construction management; we'll have survey crews on a site doing construction plans, plotting all the roads, laying out all the turbines. We'll have our materials-testing group doing compaction testing for new roads, and then concrete testing when they're pouring the turbine foundations. Then we'll do the post-environmental surveys. In most cases, we sign a three-year contract to go back every few months and do an environmental review to make sure the turbines are not negatively affecting the environment."

It is in their effect on human beings that turbines have of course incurred their greatest public resistance. "There's lots of opposition to them for sure," acknowledges Mark. "At one point, one of our geotechnical groups was confronted at gunpoint by a landowner, who threatened to shoot our people if they didn't vamoose. We also encounter a fair bit of resistance at the open houses and public forums that are required by law before these projects begin. Often there's four or five hundred people at such events, demanding answers to tough questions. These days, it's not acceptable for surveyors or engineers simply to hide out when the questions get difficult, or to run when the job's completed. As I see it, the closer you are to the forefront of a society's advances, the closer you often are to controversy; you've got to accept that and deal with it.

"I guess, overall, you could say that our work in renewables is a kind of microcosm of how the company has diversified and learned during the past dozen years. But it's also been a good lesson for us in how the culture itself is moving, and in the importance of surveyors moving with it."

Tulloch Engineering is currently at work on wind projects in Windsor and Chatham, as well as in the Sudbury, Ottawa and Niagara Falls areas – and is involved in ten solar projects across the province.

And by these marks shall the AOLS be known. From the top: The current logo of the Association, designed for the AOLS centennial in 1992. The Association's official crest, in use at least as far back as the late 1950s. The official corporate seal of the AOLS, in use from the 1930s to the present.

archivist and librarian, member of several committees, and editor/publisher of the quarterly magazine *Ontario Professional Surveyor*. Two more Erindale grads, Tim Hartley and Al Worobec, head up the Association's survey review department; and Bill Buck is the organization's registrar, a role he has held with unassuming and soft-spoken diligence for more than 16 years. "Bill and Blain are a perfect balance for one another," said an AOLS Council member in September 2016. "Whereas Blain is extroverted and outgoing, Bill is this quiet, bedrock sort of guy, who's at work every morning at seven, who seldom gets up from his desk, and who when you need him for advice or perspective, or for just plain knowledge of the organization and its issues, is an indispensible and steadying resource."

Where the Association's 505 members at large are concerned, Martin encourages professional development, as well as social engagement of a sort that in a bygone era might better have been characterized as "fellowship." He believes in gatherings, in sociability, in hospitality – and is often at the centre of such gatherings. For some forty years, he has accompanied a group of Erindale grads called the Cambrian Society on their annual winter cross-country ski trip and social fling in the highlands of central Ontario.

By his own estimate, Martin's greatest administrative challenge, ongoing, has been the yearly turnover in Association presidents. "They're very bright people," he says, "but you still have to provide them some direction, build an administrative framework for them." Which occurs in part at the Association's annual spring strategic planning session, held typically at a central Ontario resort. "The executive director is the continuity between presidents and councils," Martin says. "The planning session gives the new council a chance to see how the old council did things."

One of Martin's current initiatives is a grand conclave for the Association's 125th AGM, a gathering that for the first time will bring together the AOLS, the Ordre des arpenteurs-géomètres du Québec and the Association of Canada Lands Surveyors. The groups represent more than half the surveyors in the country. "Can you imagine?" said Martin. "Two hundred and fifty years of surveying in Upper and Lower Canada, and the surveyors of Ontario and Quebec have never gotten together!"

GIVEN THE DECLINE in the number of articling students since the dissolution of the Erindale program, the AOLS has, in recent years, conducted an exacting and energetic program in support of geomatics education and public awareness of the profession, and in the recruitment of surveyors. Barely a week goes by when Maureen Mountjoy, in her educational role, is not involved in a job fair, a trade show, in

The staff at AOLS headquarters – November 2016. Back row, left to right: Joyce Tenefrancia, Al Worobec, Tim Hartley, Lena Kassabian, Bill Buck, Herman Bernardo. Middle row: Julia Savitch, Maureen Mountjoy, Blain Martin, Sheila Lavina, Penny Anderson. Front row: Barkley. Absent: Vladimir Oppenheim.

handing out a geomatics award to an engineering student at Ryerson or York, or generally promoting the profession at a college, university or high school. During 16 years with the Association, she has also raised hundreds of thousands of dollars for the AOLS Educational Foundation. Started in 1973, the fund currently stands at nearly half a million dollars, the interest on which (combined with other support) is handed out regularly in the form of awards to geomatics and surveying students.

But advances for the contemporary surveyor and the AOLS are by no means stoked entirely by technology or education or communications. They are also the result of a whole new stratum of ethics and social forces that have required surveyors to take another look – in some cases a harder and more detailed look – at the surveying and engineering of the past; and, more importantly, at surveying's prospects for the future.

Since the surveyors themselves are best positioned to comment on what lies ahead, let us, in hastening toward the end of our book, look briefly at surveying's future and interests through their eyes.

* * *

In the view of Susan MacGregor, the profession must strive to bring its issues more productively before the government. "It's crucial, for example, that we press for renewed maintenance of the survey fabric," says the Surveyor General. "The more it deteriorates, the more difficult and costly it gets to find and refresh. In 2015, I put forward to the government the need for for re-investment, but it wasn't considered a priority at the time."

Susan will also be attempting to raise the profile of surveyors both publicly and in government circles, and to promote the enormous contribution they make to private development and to the government's ongoing creation and renewal of public infrastructure – of, among other things, highways, bridges, public transit, colleges, hospitals and courthouses.

"And as always these days," she says, "I'll be urging anybody who will listen to get behind the reestablishment of a unified survey science program at an Ontario university. Eighteen years after the fact, I still feel that the loss of the Erindale program has contributed to a lot of the issues we're facing – in particular the shortage of surveyors coming into the profession."

* * *

"If as surveyors and managers of land information we are to have any relevance in Canadian society, going forward," says Izaak de Rijcke, "we need to listen to what the courts are telling us about the importance of the First Nations' relationship to the land. We can call ourselves

AOLS Coumcil then (1894) and now (2015).

Ontario land information professionals until the cows come home, but it's just hot air if we don't attempt to embrace what it means. This year's report from the Truth and Reconciliation Commission talks about re-addressing the past, about restoration, about allowing the First Nations to re-experience and practice their own culture. Ultimately that report could be a means by which Indigenous peoples will be able to reconnect to the land. If the possibility of that happening is lost – is rejected, for example, by Canadian banks claiming they can't mortgage a piece of property if it's subject to an aboriginal title – we're not going to move forward. Novel ways have to be found to accommodate that process. As surveyors, we can help people reconsider who we are as Canadians and how we got here; can examine the role we played as a profession in what is now understood to be a great wrong. If as surveyors we can be part of a new Canadian order, that's far more significant than telling Canadian society that we're important."

* * *

"People today are the Twitter generation – they love technology and have an attention span of about six seconds," says Andy Shelp. "On the positive side, technology is attracting new people to surveying. The problem is that the only thing it provides in the end is numbers. Surveyors have to know what to do with those numbers, how to introduce them to

the broader world through a broader interpretation. As much as this is a math-based profession, it's also a law-based profession and a socially based profession. Do the numbers give you an appropriate answer to your problem or challenge? Certainly not always. Whereas surveyors used to know forestry, geology, planning, civil engineering, maybe even ethics, now they know technology. The people coming up have a great grasp of it – better, say, than mine. But they don't always have the breadth of understanding that good surveyors need. And that is what we have to get back to."

* * *

"I would say that in the next two to four years, more and more surveying will be done with terrestrial LiDAR scanning and with the resulting point clouds," says Mark Tulloch. "Right now if we're doing a lot-grading plan, which is basically just figuring out the topography on a vacant lot, our crew takes a GPS unit and, in the course of maybe an hour, ties in the entire topography for that piece of property. But there's nothing stopping us from setting up a LiDAR unit, pressing a couple of buttons and, within ten to twelve minutes, having that same lot surveyed and at a magnitude of higher accuracy and precision. You let the unit spin for a couple of minutes, and you have what you need. The reason we're not actually doing that is that the mindset of the industry hasn't caught up with the available technology – it's difficult to determine what exactly is being protected by the delay. Maybe it's a deep-down fear that technology is going too far too fast. But it's all going to come, and probably quite soon."

* * *

"Because I just got commissioned, I'm not really in a position to say anything on behalf of other surveyors," says Simon Kasprzak. "But it's pretty clear there are going to be some serious changes in the profession over the next twenty years – everything's going to get thrown on its head. And yet I feel compelled and excited to get into it – and sometimes a little nervous; five years from now, I may be selling muffins out of the back of my van. Meanwhile, if my thoughts were a blueprint for the future of surveying, I'd say the only way forward right now is to step up and embrace what's coming, and don't look back... The profession at this point is pretty much as exciting as you want to make it."

Sustaining Sponsors

This book is produced with the generous support of individual and corporate AOLS members and suppliers to the profession. Thank you to all.

PLATINUM

Callon Dietz Incorporated Ontario Land Surveyors
Ontario Ministry of Natural Resources and Forestry
Tulloch Geomatics Inc.

GOLD

Arthur J. Gallagher and Intact Insurance Company
Cansel
Rady-Pentek & Edward Surveying Ltd.

SILVER

Adam Kasprzak Surveying Ltd.
Annis O'Sullivan Vollebekk
Archibald Gray & McKay Ltd.
Coote, Hiley, Jemmett Ltd.
Douglas A. Culbert
Fairhall Moffatt & Woodland Ltd.
Krcmar Surveyors Ltd.
Leica Geosystems
Lloyd & Purcell Ltd.
Monteith & Sutherland Limited
Schaeffer Dzaldov Bennett Ltd.
Sokkia Corporation

SUPPLIERS

Tom A. Bunker
Four Point Learning
iLookAbout
Teranet Inc.
AOLS Regional Group - Hamilton & District

BRONZE

Ronald A. Denis
Robert Fenn
Vladimir Krcmar
Fred Petrich

C.T. Strongman Ltd.
Collett Surveying Ltd.
Cunningham McConnell Limited
ertl surveyors
Dearden and Stanton Ltd.
George Bracken Ltd.
Hopkins Chitty Land Surveyors Inc.
Kim Husted Surveying Ltd.
Land Survey Group Inc.
Leslie M. Higginson Surveying Ltd.
Maltman Group International
P. Salna Company Ltd.
Rugged Geomatics Inc.
Searles Surveying Ltd.
Stantec Geomatics Ltd.
Surveyors On Site Inc.
T.A. Bunker Surveying Ltd.
Tarasick McMillan Kubicki Limited
Tham Surveying Limited
The Connectors Insurance Group Ltd. - Bob Morrow
Van Harten Surveying Inc.
WSP-MMM Geomatics

AOLS Regional Group - Eastern
AOLS Regional Group - Kawartha-Haliburton
AOLS Regional Group - South Central
AOLS Regional Group - South Western

MEMBERS & SUPPORTERS

Geoff Aldworth
Drew Annable
Eric L. Ansell
Anonymous
Dino Astri
Bruce Baker
Brian Ballantyne
André P. Barrette
R. Grant Bennett
Greg Bishop
John Breese
Bill Buck
Jeff Buisman
Kenton H. Campbell
Marvin Chau
David Churchmuch
Michael Clancy
Ross A. Clarke
Vicky Culbert
Line Delorme
Izaak de Rijcke
Dan Dolliver
Denis Dutrisac
Don Edgar
F. Dale Eplett
Jeff Fee
Nicole Finn
Robert T. Force
Marc G. Fournier
Rodney Geyer
Gordon S. Good
Steve J. Gossling
Gordon Gracie
Nancy Grozelle
Edward G. Gurnett
Timothy D. Hartley
Travis Hartwick
Bob Hawkins
Jason P. Hazen
Ed Herweyer
James L. Hill
Pat Hills
Russ & Vicki Hogan
Stephen Hook
Harold Hyde
A. Gary Irwin
Ron M. Jason
Douglas Jemmett
E. Bruce Johnson
John Kennedy
Roy S. Kirkup
Goran Lale
Peter Lamb
David Lambden
Grant Lee
Murray J. LeGris
Patrick Levac
Susan F. MacGregor
Brian Maloney
Andrew Mantha
Blain Martin
Jeremy Matthews
Michael Maughan
Dan McLaren
Bruce McMurchy
Marvin and Rosemary McNabb
Paul A. Miller
W. Harland Moffatt
Peter G. Moreton
Bob Mountjoy
Maureen Mountjoy
T. Martin Nisbet
Lorraine Petzold
Ron Querubin
R. Desmond Rasch
Talson Rody
Tracy Rouse
Alister Sankey
Tom Senkus
Andy Shelp
Ralph A. Smith
Tony Smith
Chester Stanton
Jim and Annette Statham
Ronald J. Stewart
Carol Street
Peter J. Stringer
Normans Taurins
Andrea Tieman
Paul Torrance
Dan Vollebekk
Mark Thomas Watson
Brian Webster
Jason Wilband
Ken Wilkinson
Gordon H. Wood
Alan J. Worobec
George and Janet Wortman
Paul C. Wyman
Jack K. Young

Clancy & Hopkins Surveying Ltd.
Dinsmore & England Ltd
Fiddes Clipsham Inc.
Hewett and Milne Limited
Ivan B. Wallace Ontario Land Surveyor Ltd.
Lay Councillors – Miranda Paquette, Patricia Meehan, Peter Meerveld, Kathleen Gowanlock, Mark Spraggett
Logan Wealth Management
Ministry of Transportation Ontario
Telecon Design Inc.
URISA Ontario
Young & Young Surveying Inc.
Zubek, Emo, Patten & Thomsen Ltd.

AOLS Regional Group - Georgian Bay
AOLS Regional Group - North Eastern
AOLS Regional Group - Northwestern

ACKNOWLEDGEMENTS

DESPITE WHAT ITS COVER MIGHT SUGGEST, a book such as this is not the effort of a single person, or even a dozen or two dozen people, but rather a hundred, perhaps 200, most of whose welcome contributions (interviews, stories, photos) are imbedded, along with their names, on the book's pages, where readers will discover and note them.

Many more contributors have provided generous financial support, acknowledged gratefully in another part of the book.

My aim for the moment is to shine a light on those whose unheralded contributions would not otherwise be recognized. I think immediately of the excellent and accommodating staff at AOLS headquarters, who over the past year and a half have shown me warmth and hospitality, and have regularly gone beyond duty in offering me technical, archival and research assistance. For starters, I owe sincere thanks to Sheila Lavina and Herman Bernardo of the Association's review department; and to Joyce Tenefrancia at the administrative assistant's desk (and to her predecessors, Maryellen Benner and Tais Lintz); and to Vladimir Oppenheim, the Association's amiable and soft-spoken accountant, with whom I have chatted often about family and music (he is a former concert pianist). And to the office manager, Lena Kassabian, who on so many occasions has provided me with valuable perspective and advice and has invariably brightened my arrival at the office with an upbeat welcome. And to my "family" on the top floor – the matchless and industrious programs manager, Julia Savitch... and the member services coordinator, Penny Anderson, who over the months has so often brought her infectious humour and conversation to my door.

My thanks are due, too, to the organization's imperturbable registrar, Bill Buck; and to Al Worobec and Tim Hartley, two keenly knowledgable surveyors who on any number of occasions broadened my perspective during their amiable visits from the review department. It was Bill along with Bob Mountjoy who double-checked each name in the Register of Surveyors that begins on the opposite page.

Likewise, I owe a special debt of gratitude to my youthful assistant and occasional lunchmate, Narineh Kassabian, who did such a patient and resourceful job in transcribing the hundreds of hours of interviews that constitute the foundation of the book. I am equally indebted to the writer, Dawn Promislow, who painstakingly fact-checked and copyedited the text. And to Ron Stewart, Andrea Tieman and Vicky Culbert, all members of the book task force that guided the project forward.

Along the way, there were any number of surveyors who took a personal interest in the requirements of putting the book together and became unofficial adjuncts to my efforts. I am thinking of, among others, Jim Hill, Bob Gunn, Tom Bunker, Anne Cole, Bob Halliday and Jack Monteith. On visits to the Surveyor General's office, I got help well beyond the dictates of duty from Eric Ansell, Donna Gertridge and Lisa Casselman.

Meanwhile, the book would not exist except for the serious and devoted commitment of Dan Diamond, who day after day through the long weeks of production called upon his well-honed skills and judgement (not to mention his considerable humour and forbearance) in amalgamating text and illustration into the elegantly designed object that you hold in your hands.

Many thanks to Dan.

And many more to my indispensable guide and consultant, Maureen Mountjoy, the Association's deputy registrar, as well as its unofficial archivist and librarian. From the moment I began, Maureen has been my ultimate and unstinting source of information, photos, advice, proof-reading and historical perspective. I cannot overestimate my gratitude to her.

Nor can I say enough about my goodwill toward the Association's executive director, Blain Martin. Throughout, he has been a constant source of imagination, energy and creative impetus. Ever-curious about my progress in the endeavour, he has nonetheless allowed it to evolve in trust and camaraderie and freedom from interference. More than that, I could not have asked. He has my hearty and ongoing appreciation. As does everyone else who has helped us along.

My warmest good wishes to you all.

Charles Wilkins
January 2017

PHOTO CREDITS

Algoma Historical Society collection, 149; Algonquin Park Archives, 76 right and bottom; Penny Anderson, 211; courtesy Eric Ansell; Archives of Ontario, 89, 163 top; AOLS archive, back cover centre and right, ii, iii, 2, 3 left, 4 left, 6, 7, 8 left, 9, 10, 12, 13, 17, 182, 20 left, 20.2, 24-26, 28, 30 left, 32 right, 33 right, 37, 39, 45, 46, 48, top, 49, 50 left and centre, 54, 55 right, 56, 58, 59, 61, top, 66, 67 right, 68, 70, 73, 75, 76, top, 78, top and bottom right, 77, 79, 80, top left and right, centre bottom right, 81 right and bottom, 84, 86-88, 93, 96 bottom, 97, 103, 104-108, 111 bottom, 114, 115, 117, 118 bottom, 119, 121, 130, centre and top, 131 right, 133, 134, 135, 138, 139 right, 141 left, 143 left, 147, top, 150, 152, 153, 155 left, 158 bottom, 167, 178 left, 181, 185, 187, 190, 191, 192, 194, 195, 198 bottom, 205 right, 206, 207, 209, 210, 212; Brookfield Power, 208; courtesy Wayne Brubacher, 21.4; courtesy Tom Bunker, 40, 41; Tom Bunker, 184; Illona Campbell, watercolour on paper, 1984, 5; Canada Post, 112; Canada's Wonderland, 203; The Canadian Aviation Museum, 137; Canadian Cartoonist Hall of Fame, 140 bottom; Canadian War Museum, 128, 129, 151, 154, 155 right; CBC Archive, 143 right; City of Toronto Archives, 148 top; courtesy Bob Clipsham, 30 right; Cornwall Standard Freeholder, 177; courtesy Sophie-Rose Côté, 22; courtesy Crystal Cranch, 47; Bob Gunn, 3 right, 4 right, 34, 196, 197; The Don Mills Project, 140 bottom; courtesy James Ferguson, 188; courtesy Fort William Historical Park, 120; courtesy John Goltz, 182; courtesy Steve Gossling, 52; Bob Halliday, 27 bottom, 139 left, 157 left, 204, 205 left, 213; courtesy Bob Halliday, 157 right; Hastings County Historical Society, 74; Louis Helbig, 176; courtesy John Hiley, 20.1; courtesy Jim Hill, 11 top, 55 left, 57, 61 bottom, 96 top, 123, 136, 140 top, 169 top, 201; courtesy Pat Hills, 202; courtesy Mart Himma, 48 bottom; John George Howard, oil on canvas, 1833, Toronto Archives, 69; courtesy Simon Kasprzak, back flap left, 16; Ann King, 146, 156; Kitchener City Archives, 8 centre; courtesy Vaitheki Kumar, x, 21.10; courtesy Jansky Lau, 14; Laurentian University Communications, 193; courtesy Gavin Lawrence, 42; Library and Archives Canada, 111 top; LiDAR Services International, 189; Elaine M. Mackenzie, 29; courtesy Sherry Wilson McEwen, back flap right, 168; courtesy Julia Meldrum Smith, 21.6, 81 top; courtesy Jack Monteith, 53, 130 top, 180 top; Jack Monteith, 180 bottom; courtesy Peter Moreton, 162; courtesy Maureen Mountjoy, 21.5; Niagara at Large, 147 bottom; Notman Collection, McCord Museum I-39450.1, 92; Ontario Ministry of Mines and Northern Development, 85, 90, 91, 94, 95, 110, 113, 158 top, 159, 160, 161; courtesy the family of Edward Owen, 131 left; Owen Staples, watercolour, 1912, Toronto Public Library, 67 left; Devynn Penkarski, 63; courtesy Jim Purcell, 21.9, 80 top centre; Jim Purcell, 183; The Real Collectibles Shop, 166; courtesy Talson Rody, 11 bottom; St. Lawrence Seaway Authority, 174; courtesy Tham Shanmugarajah, 21.8, 43, 44; courtesy Andy Shelp, 15; Spacing Toronto Magazine, 170 top; Stratford Festival, 163, centre and bottom; Sudbury Star/Ontario Ministry of Mines and Northern Development, 178 right; www.surveyhistory.org, 230; Office of the Surveyor General of Ontario, Ministry of Natural Resources, 38, 98, 99, 141 right; courtesy Norm Sutherland, 21.3; courtesy Andrea Tieman, 186; Tracey Tong, Brockville Recorder and Times, 36; Toronto Public Library, 116; Getty Images/Toronto Star Collection – Harold Whyte, 170 bottom; courtesy Mark Tulloch, 21.7; unknown origin, 6 right, 32 left, 33 left, 35, 60, 132, 148 bottom, 173 top, 198 top; Upper Ottawa Valley Chamber of Commerce, 179; courtesy Ivan Wallace, 172, 173 bottom; Charles Wilkins, ix, 23, 27 top, 31, 51, 62, 71 bottom, 118 top; courtesy Charles Wilkins, 50; collection of Mary F. Williamson, 78 bottom left; 199; courtesy Al Worobec, back cover middle, 142.

Register of Ontario Surveyors, 1892-2016

Surveyors with Certificates of Registration are indicated as "*CR*."

A

Abdelshahid, Aziz *#1926*
Abrey, George Brockitt *#49*
Abrey, George Spencer *#367*
Acheson, Mervyn Murray *#1084*
Adams, Clare Raymond *#1394*
Adams, Kim C. *#1802*
Adams, Mark O. *CR #198*
Adamus, Donald Frank *#1187*
Afzalzada, Haron *#1961*
Agnihotri, Anil *#1772*
Ahluwalia, Sabir *#1995*
Aitken, James Duncan *#1500*
Akehurst, William John *#1901*
Akehurst, William *#1901*
Aksan, Anna M. *#1831*
Aldworth, Geoff *#1591*
Allan, John Richard *#291*
Allan, Stewart Walker *#954*
Allen, Michael Peter Joseph *#1407*
Allison, Calvin Bruce *#419*
Allman, Albert Francis *#676*
Alton, J. Mark *#1753*
Ambler, Bryan Edward Pryce *#590*
Amer, Kenneth Randall *#932*
Amin, Khairul *CR #203*
Amirnezhad, Bahram *#1976*
Andersen, Yul *#591*
Anderson, Edgar C. *CR #44*
Anderson, Frederick John *#416*
Anderson, Harold E. *CR #55*
Anderson, Herbert McEwan *#410*
Anderson, John Drummond *#275*
Anderson, Ralph Mackenzie *#435*
Anderson, William Beaumont *#337*
Andrew, James Maxwell John *#1065*
Angus, George Page *#431*
Annable, Jeffrey Drew *#1434*
Annis, George David *#1159*
Anschuetz, David Edward *#1492*
Ansell, Eric L. *#1543*
Ansley, James Douglas *#983*
Apsey, John Fletcher *#217*
Archibald, Robert William *#942*
Archibald, Samuel Wallace *#563*
Ardagh, Arthur Gowan *#381*
Aregers, Craig G. *#1869*
Armstrong, C. G. Russell *#559*
Armstrong, Christopher Maurice *#790*
Armstrong, Donald McKenzie *#835*
Arnett, Ross Wilbert *#725*
Aron, Douglas Richard *#1498*
Aron, Herman Alfred *#772*
Ashenhurst, Harold Bertram *#637*
Ashman, Douglas Ross *#1151*
Ashworth, Duncan *#1509*
Astri, Dino Robert Sirio *#1650*
Attwood, Charles Hartley *#420*
Aubrey, Peter N. *#1860*
Audas, Roy *CR #15*
Auer, Gerhard *#1501*
Avey, Edward Bartlett *CR #16*
Avis, Roger *#1525*
Aylsworth, Charles Fraser Jr. *#220*
Aylsworth, Charles Fraser Sr. *#58*
Aylsworth, John Sidney *#111*
Aylsworth, William Robert *#62*

B

Babbage, George *#906*
Babbs, Norman Walter *#1097*
Bacon, John P. *CR #197*
Baikie, John Donald *#79*
Baila, Mircea *CR #206*
Baillie, Colin S. *CR #2*
Bair, Ali J.M. *CR #17*
Baird, Alexander *#147*
Baird, Donald Wilmot *#723*
Baird, John Ainslee *#466*
Baird, Wilmot Johnston *#436*
Baker, Bruce S. *#1551*
Baker, James Clarence *#627*
Baker, Mason Herman *#390*
Baker, William Albert *#837*
Balaban, Steve *#1592*
Balfour, William Robert *#1228*
Balmer, Ross Hallett *#1081*
Bangs, Lloyd George *#800*
Bannon, Geoffrey L. *CR #155*
Barber, John Donald *#1171*
Barcham, Douglas R.P. *#1050*
Barich, Edward *#1243*
Barkosky, Richard Allan *#1382*
Barnes, John Duncan *#734*
Barrett, Russell H. *#338*
Barrette, André P. *#1763*
Barron, David A. *CR #76*
Barrow, Ernest George *#149*
Barrow, Walter Richard *#1479*
Barry, Ralph Willis *#1239*
Bartley, Thomas Holmes *#483*
Batchvarova, Tania Nenova *#1941*
Batterham, Robert J. *CR #3*
Baya, Martin *#1913*
Bazett, Edward *#179*
Beacom, Keith Irwin *#951*
Bearman, Frederick Robert *#773*
Beatson, John Campbell *#978*
Beatty, David *#107*
Beatty, Frank Weldon *#517*
Beatty, Herbert John *#286*
Beatty, John Hilliard *#572*
Beatty, Walter *#38*
Beatty, William Benjamin *#484*
Beatty, William James *#1378*
Beck, Norman *CR #83*
Becker, Gerhard Adolf *#1499*
Bedard, Mark *#1888*
Beerkens, John M. *#1771*
Belal, Walid *CR #202*
Bell, Andrew *#332*
Bell, Andrew *#95*
Bell, Bruce Jack *#1064*
Bell, Frederick Archibald *#481*
Bell, George Harry *#1333*
Bell, James Anthony *#128*
Bell, Ronald D. *CR #107*
Bellach, Guenter *#1099*
Bello, Oladele S. *#1853*
Belohlavek, Vladimir V. *CR #49*
Benedict, Paul J. *#1800*
Benedict, Ralph James *#1375*
Beninger, William Albert *#873*
Benn, Ronald *#850*
Benner, Frederick James King *#455*
Bennett, Newton D. *#584*
Bennett, Robert Grant *#1614*
Bennett, William Ernest *#1292*
Benninger, Alfred George *#699*
Beresniewicz, Chris *#1836*
Berg, Kenneth Frederick *#1252*
Berg, Ronald E. *#1737*
Beri, Raj Kumar *CR #18*
Berkeley, Garnet Lawrence *#511*
Berman, Max *#907*
Berrigan, Michael E. *CR #50*
Beverley, Malcolm George *#868*
Bews, David Matthew *#977*
Bezaire, Bernard J. *#1502*
Bhatti, Wikar A. *#1754*
Bianchi, David *#1885*
Biason, Ernest *#1184*
Biason, Lawrence *#1606*
Biddle, Gordon Arthur *#908*
Bigger, Charles Albert *#183*
Bigger, Samuel Howell *#341*
Bingham, Edwin Ralph *#360*
Bird, Sidney John Glenn *#938*
Biro, Kalman *#1202*
Bishop, Gregory C.P. *#1593*
Bishop, Harry Curry *#934*
Black, David *#1702*
Black, Douglas Hugh *#729*
Black, Sidney Wilfred *#709*
Blackburn, John Crowther *#909*
Blackburn, Phillip Ardon *#1104*
Blackie, William Vacy *#1048*
Blair, Desmond *#658*
Blair, William John *#342*
Blais, Denis *#1594*
Blake, Frank Lever *#125*
Blandy, Oliver Rowland *#437*
Bleaney, James Harry *#870*
Blyth, John McDonald *#526*
Bode, Ralph L.T.W. *#1738*
Boehme, Kerry N. *#1580*
Bogdanov, Yuriy *#1967*
Bogue, Colin Bernard *#1651*
Boileau, Murray A. *CR #184*
Bolan, William Edward *#1051*
Bolger, Francis *#76*
Bolger, Thomas Oliver *#88*
Bolton, Ellsworth Doan *#292*
Bolton, Ellsworth Doan *#326*
Bolton, Jesse Nunn *#101*
Bolton, Lewis *#84*
Bontje, Johannes B. *CR #191*
Booth, Charles Edward Stuart *#187*
Bortolussi, Adrian *#1689*
Bosco, Terry Wayne Del *#1655*
Boswell, Elias John *#305*
Boucher, Howard George *#910*
Bounsall, Andrew T. *#1861*
Bourne, Anthony C. *#782*
Bouwmeester, Ralph *#1519*
Bowden, Graham Walter *#1440*
Bowers, Francis Neil *#1565*
Bowlby, Ewart D. *CR #67*
Bowman, Arthur Meyer *#239*
Bowman, Clemens Dersteine *#165*
Bowman, Edgar Peterson *#375*
Bowman, Franklin Meyer *#285*
Bowman, Herbert Joseph *#225*
Bowman, Leander Meyer *#277*
Bowman, William James *#1278*
Bowyer, Edward W. *#1530*
Boyce, Robert Douglas *#1398*
Boyd, John Gilford *#1402*
Boyd, Robert Grant *#1208*
Boyer, Donald Frederick *#802*
Boynton, Lois R. *CR #116*
Bracken, George N. *#1760*
Bracken, George Wesley *#1054*
Bradbury, John Norris Emberson *#911*
Bradshaw, Gordon Rothwell *#580*
Bradstock, Harold Stanley *#742*
Brady, James *#66*
Brannan, Robert Edgar *#1179*
Bray, Edgar *#96*
Bray, Harry Freeman *#190*
Bray, Lennox Thompson *#333*
Bray, Samuel *#142*
Brewer, William Arthur *#1172*
Brian, Michael Edward *#361*
Brick, William Reginald *#1219*
Bridges, Ron *#1917*
Briggs, Arthur Charles *#1015*
Briggs, Thomas Edwin *#718*
Brisco, Ernest Clayton Jr. *#653*
Brodie, James Herbert *#805*
Brooke, Michael Edward *#1274*
Brooks, Thomas Harry *#1268*
Brotherhood, Rowland Wilfred *#618*
Brotherston, David Ray *#893*
Brouse, Douglas David *#970*
Brouwers, Bruce *#1553*
Brouwers, Harry Anthony *#1237*
Brown, David Benjamin *#251*
Brown, David Benjamin *#344*
Brown, David Rose *#16*
Brown, Donald Harold *#1620*
Brown, Everett Lynn *#1293*
Brown, George Laing *#311*
Brown, John Alexander Waddell *#429*
Brown, John Smith *#18*
Brown, Merrill Douglas *#1106*
Browne, Dennison Headley *#1216*
Browne, Harry John *#113*
Browne, William Albert *#133*
Browne, William Herbert *#411*
Broxham, Andrew James *#1971*
Brubacher, David M. *CR #141*
Brubacher, Wayne Douglas *#994*
Bruce, Douglas Robert *#1230*
Bruce, James *CR #84*
Brun, Jacques *#1381*
Bruneel, Roger C. *CR #117*
Brydson, Stephen A. *CR #156*
Buck, William David *#1295*

Buckle, Alan D. *CR #157*
Buisman, Jeffrey E. *#1768*
Bull, Peter M. *#1552*
Bunker, Chris *#1947*
Bunker, Thomas Alan *#1323*
Bunten, Berthold Erich *#1167*
Bunton, Kenneth Stephen *#1640*
Burchnall, Ralph Parker *#596*
Burd, James Henry *#359*
Burden, Donald Francis *#874*
Burg, W. John *#1529*
Burgess, Edward LeRoy *#350*
Burgham, Allen Russell *#813*
Burke, James Morley *#719*
Burke, William Robert *#156*
Burnet, Hugh *#231*
Burt, Frederick Percy *#211*
Burton, Ross Ivan *#1034*
Burwash, Nathaniel Alfred *#351*
Burwell, Herbert M. *#139*
Bush, Clayton Elgin *#389*
Butler, Matthew Joseph *#152*
Butterfield, John Egerton Claudon *#562*
Byrne, Thomas Henry *#402*
Byrne, Thomas *#67*

• C •

Caddy, Cyprian Francis *#54*
Caddy, Edward C. *#10*
Caddy, John St.Vincent *#97*
Cadeau, Francis M. *CR #142*
Cahill, Thomas Gordon *#992*
Calder, Alister C. *CR #32*
Callaghan, Robert P. *#1784*
Callon, Terrance O'Neil *#1017*
Calonia, Gualberto *#1982*
Cambie, Henry John *#59*
Cameron, Alfred John *#252*
Cameron, Andrew *#1314*
Cameron, James Joseph *#663*
Cameron, Thomas W. *CR #190*
Campbell, Alex Stuart *#403*
Campbell, Archibald William *#208*
Campbell, Brian R. *#1747*
Campbell, Charles Donald *#505*
Campbell, Douglas Keith *#1289*
Campbell, Elizabeth Ann *#1652*
Campbell, Gordon J. *#1510*
Campbell, James Stanley *#1192*
Campbell, John Patrick *#777*
Campbell, Kenton H. *#1810*
Campbell, Malcolm E. *CR #85*
Campbell, Norman Bruce *#1022*
Cannon, Donald Arthur *#1343*
Carbert, J. Alfred *#131*
Carbis, Walter James *#1315*
Card, Steven Joseph *#1566*
Card, William Harry *#1269*
Cardinal, Kenneth J. *CR #178*
Carlyle, Burnham Wellington *#1356*
Carmichael, William Charles *#1210*
Carmody, Joseph Archibald *#787*
Carnegie, J. Trevor *CR #109*
Caron, Marcel *CR #86*
Carpenter, Henry Stanley *#322*
Carpentier, Yvon *CR #33*
Carr, William Hamilton *#752*
Carre, Henry *#63*
Carroll, Cyrus *#50*
Carter, Edward Christian *#1157*
Carter, John Douglas *#1095*
Casgrain, Joseph Phillipe Baby *#227*
Cash, John Edward *#1392*
Cassels, Westcote Lyttleton *#485*
Cavana, Allan George *#135*
Cavana, Allan George *#238*
Cavana, Ernest Leopold *#546*
Cavell, Edward *#456*
Chaderton, Julian Cuthbert *#569*
Chalmers, Irving Kennedy *#630*
Chalmers, John *#300*
Chamberlain, Charles A.M. *CR #48*
Chambers, Donald Gordon *#1654*
Chambers, James Bertram *#622*
Chan, Phillip C. *CR #114*
Chapeskie, Raphael E. *CR #87*
Chapman, Charles Bightmer *#1004*
Chapman, Michael A. *CR #159*
Chapple, D. Brooke *#1811*
Charlesworth, Lionel Clare *#301*
Chartrand, Donat Emile *#506*
Chase, Albert Victor *#393*
Chau, Marvin Ming *#1531*
Cheesman, Thomas *#30*
Cherian, Boney *#1962*
Chidley, Murray Allan *#853*
Chipman, Willis *#181*
Chitty, Phil W. *#1886*
Chow, Sing Hon *#1347*
Christie, John Henry *#804*
Christie, Uriah Wesley *#347*
Christopher, Desmond A. *CR #160*
Church, Paul Leslie *#1466*
Churchmuch, David N. Edward *#1265*
Clancy, Michael Joseph *#1387*
Clancy, William Ronald *#1338*
Clark, John William *#1464*
Clark, W. Bruce *#1690*
Clarke, Alvin James *#912*
Clarke, Barry John *#1567*
Clarke, Carlton Henry *#1443*
Clarke, Claude James *#759*
Clarke, Frederick Fieldhouse *#348*
Clarke, Gordon Brock *#1461*
Clarke, Leonard Oswald *#339*
Clarke, Peter Thomas *#789*
Clarke, Robert Nash *#1160*
Clarke, Ross Albert *#1201*
Clement, Kim *#1460*
Clipsham, Robert Esmond *#1254*
Cloutier-Gunn, Christine M. *#1749*
Coad, Brian A. *#1781*
Coad, Richard *#167*
Cochrane, Alan Shapter *#1077*
Code, Abraham Silas *#303*
Code, Charles Edward *#515*
Code, Richard Stanley *#376*
Code, Robert Bryce *#641*
Code, Robert George *#634*
Code, Robert Wilmot *#432*
Code, Samuel Barber *#349*
Code, Thomas George *#377*
Coe, William Ronald *#902*
Coggan, George Douglas *#1414*
Colby, William David *#535*
Cole, Edward Frank Lyn *#1283*
Cole, J. Anne *#1542*
Cole, Margaret E. *CR #118*
Coleman, Richard Herbert *#150*
Collett, William Brent *#1641*
Coltham, James Thomas *#433*
Coltham, William George *#452*
Comery, David A. *#1803*
Connolly, Thomas William *#945*
Consoli, Guido Vincent *#1511*
Conway, Victor Edward *#1003*
Cook, Alfred *#537*
Cook, Beverley Graham *#1149*
Cook, Godfrey Henry *#1206*
Cook, William A. McMichael *#414*
Coons, Herbert Lindsay *#715*
Coons, Scott E. *#1788*
Coote, Basil Douglas *#1039*
Copeland, Charles David *#688*
Copeland, John Alfred *#621*
Corbett, Charles Patrick *#753*
Cormier, Dan J. *#1801*
Cornett, Sarah J. *#1691*
Cosmo, Matthew Di *#1863*
Costello, Barry W. *CR #19*
Côté, Sophie-Rose *#1987*
Cotterill, John Stanley *#1413*
Cotton, Arthur Frederick *#120*
Coulas, Timothy A. *#1608*
Coupland, Harry Gordon *#1024*
Coutts, Hugh S *#1837*
Cox, Clarence Daniel *#693*
Coyne, Paul A. *#1703*
Cozens, Joseph *#126*
Cram, Alexander Scott *#608*
Cranch, Crystal R. *#1805*
Crann, Wayne F.R. *CR #161*
Crawford, Peter Roy *#1087*
Crerar, Samuel Rutherford *#366*
Creswicke, Henry Jr. *#82*
Crewe, Richard Henry *#791*
Crocker, J. Paul *#1977*
Cromwell, Joseph Miller Oliver *#9*
Cronier, Eric M. *#1704*
Crouch, Milton Edward *#474*
Crowther, W. Seville *CR #68*
Crutchlow, Michael R. *CR #4*
Culbert, Douglas A. *#1527*
Culham, Blair *#1452*
Culham, Douglas *#1520*
Culham, Warren T. *#1581*
Cullen, Donald James *#1253*
Cummings, Dwayne *#1928*
Cunningham, Frederick G. *#950*
Cunningham, Patrick John *#913*
Curie, Lise Roxanne *#1892*
Curl, Gary Norman *#1366*
Currie, Donald M. *CR #179*
Currie, Harry Douglas Gibson *#606*
Cuthill, Ronald Robertson *#1124*
Cybulski, Daniel Alphonse *#793*
Czajka, Stephen D. *CR #132*
Czerwinski, Stefan Eugene *#1628*
Czerwinski, Tomasz *#1537*

• D •

Daley, John Francis *#1000*
Dalrymple, Douglas William *#1411*
Dalton, John Joseph *#153*
Daly, John Christopher *#1020*
D'Amico, John M.J. *#1714*
D'Amico, Tony *CR #133*
Daniels, William Joseph *#1304*
Dauphin, Edgar Leon *#757*
Davidson, Alexander *#43*
Davidson, Steven Palmer *#1939*
Davidson, Walter Stanley *#201*
Davies, Bryan Thomas *#1174*
Davies, Paul V. *CR #5*
Davies, Vernon Russell *#629*
Davis, Allan Ross *#221*
Davis, John *#157*
Davis, Kelly P. *CR #196*
Davis, William Harry *#903*
Davis, William Mahlon *#210*
Dawson, Donald K. Fortescue *#687*
Day, Nigel A.P. *#1748*
Daykin, S.E. *CR #105*
Deacon, Thomas Russ *#281*
Deane, Michael *#12*
Deans, William James *#202*
Dearden, James Dudley *#844*
Death, Arthur *#666*
de Conkey, Franklin Venn *#864*
deGraaf, Arthur *#1049*
DeGurse, Joseph *#196*
de Haan, Peter *#1739*
de Jager, Matthew *#1983*
Delorme, Line Gisele *#1630*
Delph, Frank Boothby *#1306*
De Luca, Fernando G. *#1838*
DeMorest, Richard Watson *#253*
Dempster, Hubert Orville *#404*
DenBroeder, Ross *#1878*
Denis, Ronald A. *#1692*
de Rijcke, Izaak *#1458*
DeRosa, Pier L. *#1789*
Desrochers, James Joseph *#1258*
Dew, William Oakley *#801*
Dewar, John Myles *#1026*
Dewhirst, Peter *#914*
Diamond, James Herbert *#615*
Dickson, James *#102*
Dietz, Terry Patrick *#1568*
Dillon, David John *#1374*
Dinsmore, Ivan Carl *#823*
Dixon, Howard A. *#340*
Dixon, Richard Calvin *#1478*
Dobbie, Thomas William *#31*
Dobie, James Samuel *#312*
Dodd, John Basil *#996*
Dolbear, Cecil William *#620*
Dolliver, Dan *#1521*
Domagalski, Adam *#1921*
Donaldson, Allan Bruce *#1125*
Donevan, Frederick James *#547*
Donnelly, Charles Henry *#1066*
Dore, Richard Francis *#636*
Dore, Ronald F. *#1661*
Dorland, David Sidney *#1400*
Dosen, Vladimir *#1854*
Dotterill, Christopher Ernest *#1222*
Douglas, Robert George *#1309*
Doupe, Joseph *#68*
Doxsee, Leonard Archibald *#739*
Dreger, Gerald Karl *#1226*
Drewry, William Stewart *#197*
Druery, Elmer George Guy *#1092*
Drzic, Tom *#1393*
Ducker, William A. *#188*
Dudarewicz, Czeslaw *#1310*
Dudgeon, Garry James *#1336*
Dugas, Joseph Benoit *#678*
Duncan, John Stuart *#1242*
Dunlop, R. Dean *#1726*

Dunn, Thomas Hamilton #369
Dunning, John Rowland #731
Durante, John #1437
Durrell, William Henry #592
Dutrisac, Denis #1491
Dutton, William Lawrason #599
Dynes, Richard Fforde #473
Dzaldov, Dan #1852
Dzaldov, Ophir N. #1716

E

Eade, Joseph Ronald #878
Eadie, Louis Francis #467
Eagleson, Francis Mervin #399
Earle, Wallace Sinclair #438
Eaton, Alan R. CR #88
Ebel, David Matheson #1627
Eberhardt, Dieter #1439
Eckstein, Christopher R. CR #119
Edgerton, Richard E.M. #1483
Edgerton, Thomas W.G. #1662
Edward, Paul C. #1538
Edwards, Brian Keith #1042
Edwards, George #90
Eichmanis, Eric #1251
El-Chanti, Oussama #1990
Ellins, Murray #820
Elliott, Grange Waller #712
Ellis, Douglas Stewart #457
Ellis, Henry Disney #143
Elms, Malcolm Rex #1364
Elsmore, Trevor D. CR #31
Emo, Ron James #1115
Emode, Richard E.O. CR #113
Emon, Louis Arthur #1013
Empey, John Morgan #371
Endleman, Donald William #943
Endleman, Thomas Harold #1408
England, J. Brent #1554
Eperjesi, Stefan #1470
Eplett, F. Dale #1764
Erickson, Caroline A. CR #35
Erickson, David A. CR #37
Ertl, Lawrence O. #1782
Ertl, Otto #1215
Erwin, Robert Blake #567
Esdon, William Stuart #847
Este, Auvo Olavi #588
Este, Jouko William #633
Esten, Henry Lionel #230
Evans, George #986
Evans, John Dunlop #83
Even, James #1812

F

Fair, John #122
Fairbairn, Richard Purdom #138
Fairchild, Charles Courtland #290
Fairchild, William Howard #327
Fairhall, Charles Wesley #857
Falls, Orville Mervon #536
Farley, Henry Richard #647
Farley, Sidney Edward #394
Farncomb, Alfred Ernest #294
Farncomb, Frederick William #257
Farncomb, John Elworthy #587
Farrell, King Allen #494
Farrell, Richard Warren #915
Farrow, Graham Milton #843
Farstad, John H.M. #1743
Fassel, Anthony Francis #616
Fathi, Seyed #1975
Fawcett, Milton Arnold #1014
Fawcett, Thomas #174
Fazio, Diego #1349
Fearnley, Richard E. Hatton #1117
Fee, John Jeffrey #1937
Feeley, Bernard #1123
Fencott, Robert James #1424
Fenton, John Francis #842
Fenton, William Michael #1059
Feren, Peter Raymond #1932
Ferguson, George Hendry #560
Ferguson, James E. CR #64
Ferguson, Kerry D. #1615
Ferizovic, Ken #1616
Fiddes, Zachary #1957
Filmore, Stanley H. #334
Finnie, Roderick #1575
Finos, Pier Luigi #1286
Fisher, Charles Parker #852
Fisher, Michael John #1934
Fisher, Walter Dennis #1156
Fitton, Charles Edward #163
Fitzgerald, Edward #493
Fitzgerald, James W. #343
Fitzgerald, James William Sr. #35
Fitzmaurice, Maurice William #767
Flaherty, Joseph G. #892
Flater, Frederick William #245
Flatman, Arthur James #916
Fleguel, John Beverley #1647
Fleguel, Robin L. #1828
Fleischmann, Carl Frederick #1005
Fletcher, Guy Alexander #1992
Fletcher, Stephen G. #1818
Fletcher, William Jessiman #495
Fligg, Robert A. #1555
Flim, Allard V. #1790
Flim, Henry #1006
Flint, Ivan Iliffe #557
Flint, John Gregory #987
Flook, Samuel Evert #458
Fluke, Thomas Stephen #1250
Forbrigger, Robert W. CR #38
Force, Robert Tom #1436
Ford, Allan Douglas #604
Ford, Greg #1974
Ford, William Butterton #313
Foreht, Dennis Martin #1267
Foreman, Thomas Spence #861
Forth, Paul Frederick #1311
Foster, Frederick Lucas #69
Fournier, Marc G. #1882
Fournier, Raymond CR #39
Fowler, Douglas Balfour #1047
Fowler, Robert A. CR #20
Fowlie, Albert #70
Fox, Christopher #1988
Francis, John James #61
Francis, Paul M. CR #21
Franey, Michael T. CR #77
Fraser, Charles Murray #1238
Fraser, Charles #11
Fraser, Donald Gordon #1389
Fraser, Donald John #677
Fraser, Howard Gordon #825
Fraser, Robert Roderick #894
Friel, Norman Edward #1271
Fuce, Edward Oliver #362
Fulford, Francis Bruce #1359
Fullerton, Charles Herbert #368
Fulton, Robert J. #1676
Fulton, William Joseph #521
Funk, Robert Harry #799

G

Gacser, Ernest #1138
Galati, Pasquale #1644
Galbraith, Donald Horace #854
Galbraith, John #123
Galbraith, William #195
Galejs, John #1636
Gallagher, Charles Vincent #486
Galloway, Beverly Maxwell #879
Gamble, Killaly #243
Gansen, Heinrich M. CR #40
Garbutt, Jack T. #1147
Garden, Edward R. #1727
Garden, Robert Alexander #917
Gardiner, Edward #93
Gardiner, James Neil #836
Garey, William Donald Leslie #809
Gariepy, David H. CR #95
Garrard, Gordon W. CR #89
Gaspirc, Robert James Cyril #1545
Gauthier, Richard R. #1762
Gavillier, Maurice #91
Gavin, John William #614
Gelbloom, Jaime #1808
Gelbloom, Lawrence Joseph #851
Gelleta, James L.P. #1506
Gerrits, Henry John #1450
Gervais, Jean-Marie CR #90
Geyer, Rodney H. #1718
Ghofrani, Mansour #1984
Gholami, Ali #1952
Gibbons, J. Glen CR #69
Gibbons, James #264
Gibbs, Thomas Fraser #3
Gibson, Andrew #918
Gibson, Ansell Bee #561
Gibson, Charles Wilbert #628
Gibson, Colin William George #459
Gibson, George #52
Gibson, Harold Holmes #269
Gibson, Howard Milton #1107
Gibson, James Alexander #23
Gibson, Laura E. #1819
Gibson, Morton Milne #439
Gibson, Peter Silas #39
Gibson, Silas Eldon #585
Gibson, Wilbert Silas #314
Giddy, Glenn Edward #1353
Gifford, Steven James #1625
Gilbert, Frederick Arthur #530
Gilks, Veronica D. #1677
Gill, James Richard #460
Gillies, Archibald #525
Gilliland, Thomas Brown #105
Gillon, Douglas John #296
Gilmore, Mark V. #1791
Gilmour, Robert #45
Glassford, Thomas Leonard #1332
Glenday, George #875
Goadsby, J. Morgan CR #96
Godwin, Peter John #1643
Goebelle, Hugh B. #1813
Goebelle, Norman John #576
Goff, Dennis H. CR #162
Goldman, Barry D. #1814
Golinski, Waldemar #1998
Goltz, John Francis #1185
Gondo, Thomas #1942
Gooch, Frederick Herbert #671
Good, Gordon Sinclair #1111
Goodings, William Donald #822
Goodridge, Paul G. #1663
Gordon, Charles William #965
Gorman, Michael J. #1839
Gorski, Christopher CR #78
Gossling, Steven John #1430
Gourlay, John Thomas #762
Gourlay, Robert Murray #440
Gow, Robert Eckford #1126
Graham, Derek Gibson #1288
Graham, Edward Allan #963
Graham, Howard Marshall #1132
Graham, Murray MacGillivray #973
Grander, Helmut Frank #1183
Grander, Ralph F. #1759
Grant, Gary Joel #1129
Grant, Russell Reeve #430
Grassie, Charles Andrew #548
Gray, Andrew Wellington #549
Gray, James Edward #611
Gray, John #684
Gray, Maxim Theodore #556
Graydon, Aquila Ormsby #172
Green, David #1945
Green, Rodney Sheldon #1385
Green, Thomas Daniel #203
Greenfield, Kirsten M. CR #120
Greenlees, Alexander Hunter #395
Gregoire, Paul J. #1595
Grenier, Michelle CR #91
Grenkie, Edward John #1379
Griffin, Albert Dyke #265
Griffin, George Campbell #555
Griffiths, Michael A. #1868
Grozelle, Nancy J. #1824
Gubala, Chad P. CR #163
Gunn, Robert C. #1516
Gunn, Ronald Howard #946
Gurnett, Edward Grant #1118
Gustafsson, Harry Eric #1468
Gutri, John Harvey #1465

H

Hackett, Richard George #1447
Hadfield, Colin D. #941
Hagedorn, Grover Cairns #538
Haigh, John James #583
Halford, Abraham J. B. #209
Halinen, Eero #1240
Hall, Allan Gordon #743
Hall, Charles Frederick #670
Hall, William Harvey #648
Halliday, Robert Douglas #1556
Halsall, Robert John T. #1503
Halverson, Maurice Carl #1190
Ham, Jeffrey J. CR #134
Hancock, Sydney Grenville #610
Hancock, Terrel Frederick #1325
Hanes, George Samuel #352
Hanning, Clement George #40
Hanson, Robert Thomas #1241
Haramis, J. Patrick #1713
Harper, William A. #1693
Harrington, Douglas Arthur #871
Harris-Herr, Nancy L. #1786
Harris, John Walter #98
Harris, Robert K. #1532
Harrold, Phillip Arthur #883
Harry, Kalantzakos #1922
Hart, Milner #74
Hartley, Timothy D. #1528
Hartt, James Paul #949
Hartwick, G. Travis #1847
Hartwick, Gregory J. #1705

Register of Ontario Surveyors, 1892-2016 continued

Harvey, George Carlyle *#749*
Harvey, John Michael *#840*
Harvey, Thomas Alexander *#287*
Harwood, Gordon Walter *#1102*
Haskins, William *#24*
Hatkoski, Richard Jack *#1354*
Hawkins, Robert Charles *#1406*
Hawley, David J. *#1761*
Hay, Colin Clifford Robert *#1031*
Haynes, Basil Joseph *#785*
Hazen, Jason P.E. *#1880*
Head, Howard Bell *#691*
Heaman, John Andrew *#307*
Heggie, Robert Douglas *#1415*
Heinbuch, Wallace R. *#884*
Henderson, Eder Eli *#233*
Heney, Peter J. *CR #14*
Henrickson, David R. *CR #135*
Henry, Frederick *#234*
Herman, Zoltan *#1930*
Hermanson, Glenn Donald *#1058*
Hermon, Ernest Bolton *#215*
Hermon, Royal Wilkinson *#36*
Herweyer, Edward H. *#1576*
Hewett, Maurice *#654*
Hewlett, James A. *#1899*
Hewson, Thomas Ringwood *#145*
Heywood, Albert Henry *#1177*
Heywood, Allan John *#1621*
Hickson, Gerald G. *#1720*
Hietala, Johannes Ensio *#1009*
Higgin, Valerie I. *CR #164*
Higginson, Leslie *#1596*
Hiley, John Robert *#818*
Hiley, John William *#1494*
Hill, Charles McDougall *#740*
Hill, James Leslie *#1078*
Hillis, Kerry Francis *#1634*
Himma, Mart Hindrek *#1631*
Hinchley, Robert J. *CR #41*
Hobson, Joseph *#25*
Hodgson, Shawn *#1919*
Hofmann, Phillip *#1533*
Hogan, J. Russell *#1617*
Hogarth, George *#441*
Holcroft, Herbert Spencer *#335*
Holder, Robert George *#961*
Holding, Paul Tress *#1075*
Holland, William Hugh *#378*
Holm, Juergen Kurt *#1045*
Holman, Kenneth R. *CR #108*
Holstead, Donald *#657*
Homer, Peter J. *#1750*
Hook, George Stephen *#1007*
Hook, Stephen D. *#1815*
Hopkins, Marshall Willard *#288*
Hopkins, Peter Stanford *#1376*
Hoppe, Thomas *#1773*
Horton, Graydon Thring *#765*
Horwood, David M. *CR #144*
Horwood, David Orlin *#1128*
Houghton, Donald Ira *#741*
Houghton, Ward *#1706*
Howard, Charles John Benedictus *#461*
Howden, Harold Stewart *#607*
Howe, David Joseph *#706*
Howe, John Parnell *#523*
Howe, Kerry Thomas *#735*
Howes, James Andrew *#690*
Howitt, Alfred *#27*
Hu, Yahui *#1958*
Hubert, Joseph Wayne *#1266*
Huch, Hans Karl *#1127*
Hudson, George Herbert *#702*
Huffman, Karl *#475*
Hughes, William James *#919*
Hulme, Kenneth S. *#955*
Hume, Darrell Lawrence *#1360*
Humphries, David Thomas *#673*
Hundt, Aphonsus D'Arcy *#872*
Hunt, Douglas E. *#1534*
Hunt, Henry John *#736*
Hunt, John James Harmer *#1033*
Hunter, Raymond Eric *#798*
Huppert, George Leo *#1419*
Husted, Kimberly S. *#1582*
Hutcheon, James *#271*
Hutt, Thomas F. *CR #93*
Hwang, James Shao-Ming *#1316*
Hyde, Harold D. *#1827*

• I •

Iarocci, Mario *#1150*
Iavicoli, Bruno *#1832*
Ims, Theodor H. *#1797*
Inglis, Robert D. *CR #97*
Ingraham, Terry Joseph *#1285*
Ingram, Albert Cecil *#603*
Innes, William Livingstone *#278*
Ireson, Edward Taylor *#496*
Irwin, Arthur Gary *#1573*
Irwin, Bruce C. *#1728*
Irwin, James Moore *#71*
Isaacs, Ian K. *CR #42*
Isip, Reynaldo L. *#1897*

• J •

Jackson, Alan Mair *#448*
Jackson, Brian *CR #22*
Jackson, Ian Maurice *#795*
Jackson, John Edward *#1086*
Jackson, John Edwin *#421*
Jackson, John Herbert *#331*
Jackson, Lloyd Derwent *#859*
Jackson, Percival Anthony *#476*
Jackson, William *#877*
Jacobs, Bryan Andrew *#1629*
Jacques, Suzanne *#1859*
James, Darrell Denman *#270*
James, Silas *#41*
Jaros, Ronald *CR #187*
Jason, Ronald McLean *#1425*
Jayson, Paul *#1121*
Jeffray, Angela *#1927*
Jeffrey, Alexander *#660*
Jemmett, Douglas *#1550*
Jemmett, Shawn Alexander *#1648*
Jenkins, Garth Kevin *#1574*
Jephson, Richard Jermy *#144*
Jepson, William Charles *#543*
Jeraj, Alnashir *#1864*
Jewitt, Harold Victor *#700*
Jireada, Charles Fernand *#1432*
Jiwani, Zulfikar Alibhai *#1646*
Jobe, William Lawrence *#1002*
John, Scott S. *#1895*
Johnson, Edward Bruce *#1571*
Johnson, James W. *#1889*
Johnson, Robert Thornton *#254*
Johnson, Ross Melville *#1262*
Johnson, Sydney Munnings *#297*
Johnston, Herbert *#346*
Johnston, Kerry S. *#1688*
Johnston, Robert Thornton *#325*
Johnston, William James *#418*
Johnston, William James *#442*
Johnston, William James *#1116*
Jones, Charles Albert *#178*
Jones, Clare William Allan *#976*
Jones, Darrell W. *CR #128*
Jones, George Samuel *#396*
Jones, Gordon Francis *#1233*
Jones, John Henry *#226*
Jones, John Henry *#77*
Jones, Russell Wm. Roderick *#1282*
Jones, Thomas Henry *#160*
Jones, Tom Dixon *#1950*
Jones, Tudor Preece *#1016*
Jordan, Robert James *#1626*
Jordens, Douglas F. *#1619*
Julian, Terrance Michael *#1372*
Jupp, Albert Ernest *#422*

• K •

Kaczmarek, Rafal *#1955*
Kains, Tom *#118*
Kaldeway, Martin Herman *#1249*
Kapoor, Darshan C. *CR #23*
Kaposi, Bela *#993*
Karklins, Aldis *#1632*
Karpiel, Ronald Stanley *#1449*
Kasprzak, Adam *#1557*
Kasprzak, Simon A. *#1985*
Keat, John C.G. *#1678*
Keating, John Webb *#1153*
Keatley, Gordon R. *#1883*
Keatley, William James *#1298*
Keay, William Logan *#807*
Keefer, Thomas Coltrin *#2*
Keffer, Howard Franklin *#696*
Kelly, Kevin M. *CR #7*
Kelly, Michael Norris *#1320*
Kelly, Thomas Whittal *#665*
Kenea, Nasir H. *CR #165*
Kennedy, James Henry *#235*
Kennedy, James J. *#931*
Kennedy, John Hugh *#1442*
Kenney, Carmen Rice *#472*
Kerr, Brian William *#1352*
Kerr, Ernest William *#1277*
Kerr, John Henry *#935*
Ketchum, Kenneth J. *#1577*
Kiar, Christian Rudolph *#1322*
Kidd, Paul John *#1609*
Kieffer, Thomas John *#925*
Kihl, John Eric *#1152*
Kikas, Anton *#1280*
Kilkenny, John Murray *#595*
King, Adam *#1972*
King, John Albert Shirley *#527*
Kingston, Laura A. *CR #115*
Kingston, Robert Joseph *#1207*
Kinnear, Louis Arthur *#468*
Kippax, Hargreaves *#148*
Kirk, John Albert *#146*
Kirk, Joseph Green *#5*
Kirkland, James Edward *#1429*
Kirkpatrick, George Brownly *#72*
Kirkpatrick, Richard Walter *#1030*
Kirkpatrick, Roy C. *#738*
Kirkup, John Clifford *#764*
Kirkup, Roy Stanley *#1639*
Kirkup, Roy Stanley *#480*
Kirstine, Burns Gary *#1299*
Kisway, John Joseph *#717*
Kitchen, Paul Robert *#1390*
Kitto, Franklin Hugo *#602*
Kliaman, Cindy S. *#1607*
Klinkenberg, Brian *#1487*
Klinkenberg, Hans *#761*
Klotz, Otto Julius *#129*
Knisley, Martin William *#1649*
Knowles, John P. *CR #121*
Koester, Hans *#1072*
Konga, Jury *CR #60*
Koning, Jurjen Jan *#997*
Koolhaas, Dirk *#1275*
Kostic, Milija Zivko *#979*
Koudys, Albert J. *CR #65*
Kovacs, David A. *#1851*
Kowalenko, Walter *#1488*
Krcmar, Maja *#1774*
Krcmar, Saša *#1775*
Krcmar, Tomislav *#1900*
Krcmar, Vladimir *#1370*
Krebs, Harry *#1044*
Kreze, Daniel *#1622*
Kristjanson, Tom *#1722*
Krupicz, Joseph Alexander *#1401*
Krupowicz, Richard Robert *#1423*
Kruys, Robert *#1522*
Kubicki, Borys D. *#1865*
Kuelling, Laurence J. *#1564*
Kujala, Kevin P. *#1848*
Kumar, Mahadave B.C. *CR #166*
Kumar, Vaitheki *#1986*
Kupferschmidt, Martin *#1368*

• L •

Lackstrum, Eric Johannes *#771*
Ladines, Jayson F. *#1956*
Lafontaine, Edward O. *#940*
Laframboise, Gabriel *#1898*
Laird, James Stewart *#103*
Laird, Robert *#240*
Lale, Goran *#1951*
Lamb, Peter B. *#1729*
Lambden, David Whitfield *#821*
Lamont, David A. *#1829*
Lancaster, Edward *#1547*
Lancaster, Jonathan Rowland *#956*
Lane, Andrew *#293*
Lane, Douglas Arthur *#895*
Lane, Frederick Carleton *#443*
Lane, Frederick George *#957*
Lane, Lenox Thompson *#635*
Lane, Raymond Thompson *#707*
Lang, John Leiper *#384*
Lanning, John *#522*
Lansky, Theodore Edward *#988*
Lanthier, Joseph Edward *#860*
Lanzon, Silvio *#544*
LaPointe, Stéphane *#1918*
LaPrise, Joseph Michael *#701*
Larke, John Raymond *#839*
Larocque, R. Brent *#1798*
Larocque, Richard *#1257*
Lashley, Keith Floyd *#1142*
Lau, Francis *#1914*
Lau, Jansky Tak Choi *#1953*

Lau, Kin-Man *#1327*
Launen, Kari Juhani *#1324*
Lawe, Henry *#55*
Lawlor, Michael J. *#1610*
Lawrence, Charles A. Rutler *#529*
Lawrence, Gavin Eldred *#1906*
Lawryshyn, Roman Bohdan *#1067*
Laws, James M. *#1792*
Lawson, Horace Hetherington *#593*
leMay, Tracy Deavin *#400*
Lee, Robert B. *#936*
Lee, Roger Melville *#413*
Leeper, Robert Patrick *#832*
Lees, Richard Hazen *#958*
Leese, Julie F. *CR #180*
Legat, Jaro A. *#1809*
LeGris, Murray James *#1367*
Legros, Leo Arthur *#1198*
LeGrow, Neil A. *#1755*
Leiper, Rob Colin *#1997*
Leitch, John MacKinnon *#639*
Leitch, John Strickland *#487*
Lemmetty, Anita I. *#1896*
Lemon, Marvin R. *#609*
Lemon, Robert Lloyd *#763*
Lendrum, Robert Watt *#119*
Leslie, Craig *#1694*
Leslie, Jamie William *#1940*
Levac, Patrick *#1989*
Lever, David Manly *#1297*
Lewis, John Bower *#198*
Lewis, Noble Curtis *#1199*
Lewis, Stuart *#929*
Li, Amy Kwok Ying *CR #205*
Li, Jun *CR #145*
Li, Songnian *CR #167*
Lin, Joseph *#1830*
Linders, Donald J. *CR #150*
Linhares, J. Eduardo *#1825*
Lise, Arthur J. *#1963*
Little, Alan J. *CR #168*
Livingstone, Thomas Chisholm *#46*
Lloyd, Charles Winford *#679*
Lloyd, Norval Clarence *#444*
Lo, George C.M. *#1664*
Loai, Amar *#1991*
Loewen, George *#1148*
Logan, Ronald Albert *#1273*
Loginsky, Lubomyr Sergius *#880*
Lonergan, John Alexis *#623*
Lord, Rodney D. *#1679*
Lorentz, Richard Dennis *#1399*
Lorentz, Walter W. *#947*
Lougheed, Aaron *#249*
Low, Benson A. *CR #43*
Low, Edward Hamilton *#336*
Low, Nathaniel Edward *#32*
Lowe, George Merton *#980*
Lumb, William Ewart *#501*
Lumsden, Hugh David *#89*
Lymer, Daniel J. *#1597*
Lynch-Staunton, Francis H. *#33*
Lynch, Bernard Edward *#1133*
Lynch, Brian Jerome *#1642*
Lyndon, Norman Phillip *#812*
Lyon, Charles Richard *#550*
Lyon, David A. *#1598*
Lyon, John Edward *#524*
Lyons, Earl Charles *#605*
Lyons, Thomas Edward *#989*
Lyszkiewicz, Andrew A. *CR #56*

• *M* •

MacArthur, James Kenneth *#624*
MacAulay, George Beveridge *#697*
MacCallum, James *#15*
MacDonald, Charles Alfred *#1073*
MacDonald, Christopher A. *#1849*
MacDonald, Douglas Keith *#1261*
MacDonald, James Atwood *#512*
MacDonald, Thomas G. *#1822*
Macdonell, Donald John *#1023*
MacDougall, Allan Hay *#47*
MacDougall, Stuart L. *#1062*
Macek, Michael *#1605*
MacGregor, James Martin *#445*
MacGregor, Susan F. *#1656*
MacIntosh, James Alexander *#1459*
MacKay, Alexander MacLean *#828*
Mackay, Ernest George *#462*
Mackay, Gordon Fraser *#1162*
Mackay, James John *#321*
Mackay, Owen *#229*
MacKenzie, Andrew Miles *#1112*
MacKenzie, Hunter *#730*
MacKenzie, Richard A. *#1665*
MacKenzie, William Lyon *#236*
Mackenzie, William *#299*
Mackey, Robert William *#1063*
Macklin, Harold Lawson *#746*
MacLean, Douglas Charles *#1100*
MacLean, William James *#816*
MacLeod, Alistar Martin *#1489*
MacLeod, Henry Augustus F. *#34*
MacLeod, Kenneth E. *#1657*
MacLeod, Malcolm H. *CR #29*
MacMillan, James Alexander *#140*
MacMillan, John Donald *#1246*
MacMillan, Ronald George *#1029*
MacNabb, Bruce Baird *#1101*
MacPherson, Duncan *#199*
Macquarrie, Edison Malcolm *#565*
MacRostie, Norman Barry *#477*
Madan, Raj Paul *#1388*
Magee, Douglas Everett *#1181*
Magee, G. Bret *#1816*
Magnanelli, Alfio M. *CR #98*
Magrath, Charles Alexander *#182*
Magrath, Charles Alexander *#38*
Maher, Murray Ralph Willis *#984*
Mailhot-Aron, Anne-Marie *CR #99*
Main, George Anderson *#589*
Mak, Ronald M. *#1785*
Mak, Rudy *#1546*
Malcolm, Sherman Morgan *#44*
Malcolm, William Lindsay *#446*
Malcolmson, Walter S. *#469*
Maleszyk, Alexander *#1667*
Maloney, Brian *#1549*
Manigault, William Mazyck *#136*
Mann, Robert J. *#1668*
Manol, Steven *#1021*
Manouchehr, Mirzakhanlou *#1923*
Manser, Charles J. *#498*
Mansfield, Peter *#1535*
Mantha, Andrew S. *#1744*
Mantle, John Fleming *#1105*
Manton, Lewis Goodwin *#520*
Marck, Joseph Albert *#488*
Mares, Viorel *#1924*
Marion, David H. *CR #30*
Marlatt, Michael Edward *#1540*
Marlow, Robert M. *CR #70*
Marr, Douglas Gordon *#1337*
Marshall, Eugene P. *CR #8*
Marshall, George Lofthouse *#766*
Marshall, James *#99*
Marshall, Joseph *#531*
Marshall, Oscar Joseph *#720*
Marshall, William A. *#944*
Martin, Blain W. *#1454, CR #149*
Martin, Hugh J. *#948*
Martin, Malcolm Ellis *#686*
Martin, Robert C. *#1745*
Marton, Alexandru *#1907*
Mascoe, William Arnold *#1339*
Maskell, Barry Robert *#1391*
Matheson, Alexander *#519*
Matheson, Hugh R. *#499*
Mathewson, Stanley A.L.J. *CR #169*
Matthews, Jeremy C.E. *#1881*
Matthews, Kenneth Malcolm *#1272*
Matthews, Michael F. *#1740*
Matthews, Raymond John *#744*
Maughan, David U. *#1884*
Maughan, Lawrence Ulysses *#981*
Maughan, Michael J. McDougall *#920*
Maughan, Russell Gordon *#675*
Mauro, Frank *#1548*
Maynard, Donald Kenneth *#1025*
Mayo, Roy C. *#1756*
McAlpine, Michael Joseph *#776*
McAree, John *#104*
McAuslan, Herbert James *#365*
McBain, Richard Henry *#819*
McBurney, Robert Grant *#1433*
McCartney, Robert William *#1196*
McCausland, Alvin D. *CR #181*
McCaw, Robert Daniel *#372*
McColl, Charles Ross *#398*
McConnell, Kenneth Harvey *#760*
McConnell, Robert *#1724*
McConnell, William D. *CR #51*
McCuaig, Robert E.D. *CR #100*
McCubbin, George Albert *#298*
McCulloch, James Archibald *#1422*
McCullough, Andrew Lake *#247*
McCurdy, Robert T. *#1046*
McCutcheon, Vernon Spencer *#768*
McDermott, Robert M. *#1730*
McDonell, Augustine *#75*
McDougall, Samuel Gladstone *#478*
McDowall, Robert *#266*
McDowell, William Charlton *#689*
McElligott, John Anthony *#1653*
McElravy, Gordon D. *CR #101*
McEvoy, Henry Robinson *#127*
McEwen, Alexander Campbell *#784*
McFadden, Moses *#37*
McFarlane, George P. *#1485, CR #204*
McFarlen, George Walter *#259*
McGarry, Patrick Joseph *#516*
McGeorge, David Graham *#886*
McGeorge, Donald Dolsen *#985*
McGeorge, Margo L. *#1707*
McGeorge, William Graham *#423*
McGeorge, William Graham *#92*
McGrandle, Hugh *#193*
McGregor, Francis Millar *#995*
McGuinness, Peter Joseph *#1158*
McGuire, Conor *#1981*
McGuire, Gordon D. *#1751*
McHugh, Ignatius *#1154*
McIntosh, Donald Strader *#817*
McIntosh, William *#669*
McKay, Cameron Melville *#1108*
McKay, John Alexander *#881*
McKay, Robert Binnie *#685*
McKay, Scott A. *#1583*
McKechnie, Michael *#1949*
McKechnie, Stewart D. *#1508*
McKenna, John Joseph *#53*
McKergow, Frank Chester *#661*
McKibbon, Robert W. *#1708*
McKibbon, Ronald Gladstone *#1137*
McKinley, Vincent Ignatius *#786*
McKnight, George Arthur *#659*
McKnight, James Henry *#514*
McLaren, Alexander Taylor *#824*
McLaren, Ian Charles *#1335*
McLaren, S. Daniel *#1709*
McLatchie, John *#78*
McLean, Donald Peter *#865*
McLean, James Keachie *#132*
McLean, Keith Alexander *#862*
McLean, William Arthur *#315*
McLennan, Alexander B. *#638*
McLennan, Edward T. *#1093*
McLennan, Murdoch John *#289*
McLennan, Roderick *#8*
McLeod, J. Daniel *#1741*
McMaster, Douglas Gilbert *#1170*
McMeekin, Albert *#424*
McMillan, John *#1517*
McMorran, Douglas Scott *#1874*
McMullen, William Ernest *#280*
McMurchy, Bruce Ian *#1109*
McNab, John Duncan *#168*
McNabb, John Chisholm *#170*
McNabb, Marvin D. *#1558*
McNaughton, Alexander Lorne *#358*
McNaughton, Findlay Donald *#323*
McNeely, Douglas James *#939*
McNeice, Leonard Melville *#695*
McNeil, Don Wilbert John *#1245*
McNeil, Trevor D.A. *#1840*
McPherson, Archibald John *#309*
McPherson, Bruce G. *#1584*
McPherson, Charles Wilfred *#320*
McPhillips, George *#212*
McRae, Reuben *#1966*
McRostie, Gordon Callander *#632*
McSkimming, John *#810*
Meader, Charles Herbert *#509*
Meadows, William Walter *#316*
Medema, Eric Jack *#1276*
Medendorp, John *CR #24*
Medley, Graham Bingham *#732*
Medley, Stewart Dunbar *#1089*
Meisner, Robert Jeffrey *#1212*
Meldrum, Hazen Bronson *#959*
Meldrum, Julia M. *#1780*
Mellish, Herbert Leslie *#1328*
Melson, John Waller *#803*
Menzies, John Ross *#582*
Mercer, Samuel Stothard *#845*
Merrick, William George *#1041*
Merriman, Thomas Edward *#698*
Merrles, John *#1903*
Merry, William Irven *#1559*
Metz, John *#1114*
Michael, John H. *CR #110*
Middleton, John Arnold *#1085*
Middleton, Ronald Stanley *#708*
Milbrath, James W. *CR #62*
Miles, Charles Falconer *#65*
Milkintas, Kestutis J. *CR #102*
Miller, Frederick Fraser *#205*
Miller, Gordon Charteris Hire *#1481*
Miller, Homer Grant *#991*

Register of Ontario Surveyors, 1892-2016 continued

Miller, Lawrence Andrew *#1146*
Miller, Paul A. *#1512*
Miller, Richard D. *#1585*
Millesse, Leonardus C. J. *#1209*
Milne, John Campbell *#888*
Milne, Neil C. *#1855*
Milne, William R. *CR #59*
Milton, Peter Raymond *#1098*
Milway, Thomas Evan *#751*
Minnes, James Arnott *#1120*
Minnie, Steven J. *#1710*
Miret, Dario A. *#1806*
Misofi, Lambro *CR #153*
Mitrev, Simeon E *#1946*
Mitsche, Helmut *#1578*
Mo, Jason Chun-Ho *#1980*
Moeller, Robert W. *#841*
Moffatt, William Harland *#889*
Molloy, Perry A. *#1681*
Monaghan, John Wm. Lawrence *#792*
Monaghan, Patrick Anthony *#814*
Monahan, Ilona C. *CR #92*
Monteith, James Clarence *#558*
Monteith, John Dunsmore *#1053*
Montgomery, Royal Harp *#353*
Mooney, Lincoln *#513*
Moore, Edgar Lawrence *#504*
Moore, John Celeste *#882*
Moore, John Harrison *#260*
Moore, John MacKenzie *#169*
Moore, Raymond A. *#1793*
Moore, Raymond E. *CR #1*
Moore, Thomas Alexander *#282*
Moore, William James *#1623*
Moore, William James *#382*
Moore, William James *#574*
Mora, Rick Della *#1876*
Moreton, Peter Geoffrey *#1317*
Morgan, Charles Robert *#876*
Morgan, Robert Elgin *#1074*
Morris, Alfred Edmund *#164*
Morris, James Lewis *#223*
Mortimer, Richard A. Hyde *#1294*
Morton, David A. *#1826*
Motton, Charles William *#575*
Moulton, Robert L. *CR #61*
Mountain, George Alphonse *#200*
Mountjoy, Maureen Vivian *#1467*
Mountjoy, Robert G. *#1746*
Moyles, Norman Wilfred *#781*
Mrstik, Paul F. *CR #45*
Mucklestone, Francis Herbert *#479*
Mucklestone, John Kenneth *#869*
Mucklestone, Raymond Frank *#613*
Muir, John W. *#1779*
Mulcahy, Robert Rolland *#705*
Mulder, Berend *#1350*
Mullally, James Peter *#1358*
Munro, John Vicar *#295*
Murakami, John M. *CR #75*
Murdie, William Campbell *#552*
Murdoch, Robert M. *CR #136*
Murdoch, William *#51*
Murphy, Anthony D. *#1723*
Murphy, Charles Joseph *#224*
Murphy, Denis *#1079*
Murray, Kenneth Glen *#1455*
Murray, Richard Wayne *#1341*
Musclow, Chris *#1912*
Mwinyi, Omari Bakari Seleho *#1658*
Myers, Lloyd Nelson *#808*

• N •

Nanfara, Joseph *#1870*
Naraine, Robert *CR #71*
Nash, Abram Leland Stanley *#492*
Nash, Thomas Webb *#21*
Neelands, Ernest Wesley *#391*
Nesbit, Lawrence Michael *#1560*
Neville, Everett Arthur *#510*
Newlands, Jose Joseph *#1018*
Newman, John James *#317*
Newman, William *#283*
Ney, Cecil Herman *#553*
Ng, Foo Yip *#1871*
Nicholson, James Wesley *#1094*
Nicholson, Stephen Bailey *#1232*
Nickerson, Burnley Arthur *#1263*
Nicol, James *#1959*
Niculae, Roxana *#1833*
Nielsen, Peter M. *CR #199*
Nisbet, James D. *#926*
Nisbet, T. Martin *#1682*
Nisioiu, Tudor *#1908*
Niven, Alexander *#48*
Niven, David Alexander *#463*
Noble, George van de Vorde *#601*
Norgrove, David James *#1373*
Norris, Donald Earl *#1145*
Norrish, William Henry *#489*
Nouwens, John Peter *#1497*
Nouwens, Marcus J.T. *#1873*

• O •

O'Connell, Gerald Jack *#1346*
O'Connor, Shawn M. *#1867*
O'Dale, Charles Peter *#650*
O'Donnell, John Hugh *#1420*
Offer, William Neil *#891*
Ogilvie, Donald William *#1010*
Ogilvie, John Henry *#134*
Ogilvie, William *#108*
O'Hara, Walter Francis *#279*
Okoronkwo, I. Victor *CR #143*
Olafson, Harald *#933*
Ord, Lewis Redmond *#206*
Oren, Nedim *CR #208*
Orendorff, John Roche *#668*
Orr, Andrew Kenneth *#1329*
Osborne, Alexander C. *#1526*
Osinski, Marek *#1893*
Ostapiak, Daniel *#1313*
Osuchowski, Zofia *CR #200*
O'Sullivan, Michael Jefferys *#1169*
Outram, Alfred Allan *#598*
Oyler, Christopher John *#1936*

• P •

Packowski, Thomas Joseph *#1572*
Padovan, Romano *#1507*
Page, Dagmar (Dasha) *#1834*
Paice, Cyril Wilfred *#770*
Palladino, Julius *CR #201*
Palladino, Julius *#1742*
Palleck, Henry Francis *#672*
Palu, Sergi *#1122*
Panting, Stephen Bernard *#831*
Papa, Guido *#1345*
Papa, Valerio G. *#1909*
Papp, Kalman *#1383*
Parker, Anthony H. *CR #188*
Parker, Bruce A. *#1721*
Parker, Gary Lloyd *#1305*
Parker, Lloyd Edward *#815*
Parkin, Margaret J. *CR #195*
Parr, Robert Barry *#1182*
Parsons, Alison M. *#1767*
Parsons, Johnston L. Rowlett *#354*
Parsons, William Ernest *#1569*
Pastushak, Isadore Michael *#1136*
Paterson, James Allison *#158*
Patten, Lynn Harvey *#1410*
Patten, Thaddeus James *#194*
Patterson, Douglas Walter *#1290*
Patterson, Frank Elliott *#397*
Pattison, William James *#811*
Paul, Ralph William *#1480*
Paulin, Frederick William *#386*
Pavlin, Peter *#1103*
Payette, Jean Guy *#990*
Payette, Marc P. *#1778*
Pearce, Frederick John Sydney *#674*
Pearce, Wayne T. *#1669*
Pearce, William *#114*
Pearson, Michéle M. *#1680*
Pearson, Robert G. *#1670*
Peat, Christopher *#680*
Peckover, Horace Joseph *#405*
Pedder, James Robert *#272*
Pequegnat, Marcel *#406*
Perera, Wickramage Sunil *#1994*
Perkins, Kevin D. *#1695*
Perkins, Stephen M. *CR #112*
Perrie, William Wallace *#507*
Persaud, George Michael *#1638*
Pesce, David J. *#1776*
Peters, John David *#1213*
Peters, John T. *#655*
Peters, Lloyd *#1008*
Peterson, Peter Alexander *#73*
Petrich, Fred *#1536*
Petrie, Martha L. *#1701*
Petrovic, Djordje *#1970*
Pettit, D. Bruce *#1787*
Petzold, Eathel Welbanks *#830*
Phillips, Edwin Percy Argall *#407*
Phillips, Gary W. *#1586*
Phillips, John Dunlop *#966*
Phillips, Malcolm Charles *#855*
Pickard, Barry Roger *#1363*
Pierce, Benjamin Clifford *#490*
Pierce, John Gourlay *#642*
Pierce, John Wesley *#392*
Piller, Helmut *#1217*
Pilling, John Turton *#901*
Pinhey, Charles Herbert *#250*
Piraino, John P. *CR #171*
Plaxton, William James *#1161*
Poletto, Darko I. *CR #172*
Ponton, Archibald William *#171*
Poot, Robin W.L. *CR #130*
Popa, Dacian Nicolae *#1973*
Popa, Dorin *#1891*
Pope, Robert Tyndall *#124*
Porkolab, William Andrew *#1291*
Porter, Donald G. *CR #9*
Pottage, John Elder *#1301*
Potten, Harold A. *CR #26*
Poudrier, Jean-Yves *CR #111*
Power, K. Michael *CR #173*
Poyser, Beecher Doran *#829*
Praskey, Victor Robert Orville *#797*
Preiss, Richard A. *#1539*
Preston, Gary L. *#1752*
Preston, Ronald Keith *#1351*
Probert, William Joseph *#1405*
Proudfoot, Hart William *#355*
Proudfoot, Hume Blake *#184*
Prystanski, John *#774*
Pu, Tony *#1993*
Pun, Yip Kai *#1421*
Purcell, James Murray *#969*
Purcell, T. Murray *#1683*
Purser, Ralph Clinton *#534*
Purvis, Frank *#121*

• Q •

Querubin, Ron *#1965*
Quesnel, Paul Maurice *#1637*
Quinlan, Danny P. *#1579*
Quinsey, William John *#897*

• R •

Rabideau, Randall C. *#1496*
Rabin, David Aaron Raphael *#704*
Rady-Pentek, Joseph *#1318*
Raiend, Valdek *#1027*
Raikes, Peter T. *#1841*
Raikes, Rodney Campbell *#1255*
Rainboth, Edward Joseph *#241*
Rainboth, George Charles *#106*
Rainboth, George Louis *#470*
Raithby, David J. *#1684*
Ramachandran, Piratheepan *#1968*
Ramchuram, Stivell *CR #10*
Ramsamooj, Sase Naraine *#1561*
Ramsey, Guy Lawrence *#464*
Randall, James Derek *#1224*
Ransom, John Thomas *#425*
Ransom, Thomas Saunders *#1076*
Rappich, Arthur Friedrich *#1221*
Rasch, Ronald Desmond *#1321*
Rathnayake, Vineetha S. *#1943*
Rattos, Alban Nicholas *#1462*
Ratz, William Douglas *#726*
Rawlinson, Stephen G. *CR #194*
Ray, Gordon A. *#1731*
Raychyba, Edward John *#971*
Raymond, Brian John *#1256*
Reddy, Charles Francis *#1244*
Redekopp, Benjamin Peter *#1037*
Redmond, Donald Alexander *#1342*
Reed, Thomas R. *#1872*
Reiach, Lindsay *CR #123*
Reid, James Hales *#56*
Reid, John Lestock *#109*
Reid, Rodger J. *CR #47*
Reid, Rodger J. *#1766*
Reid, William Frederick *#1173*
Reiffenstein, James Henry *#115*
Reilly, William Robinson *#175*
Reinhardt, Carl *#324*
Reinthaler, Harold F. *#1666*
Reinthaler, Hubert Johann *#999*
Reis, Kuldar *#1270*
Reitsma, Peter Douglas *#1495*
Renaud, Marcel Edgar *#1474*
Rentis, George *#1264*
Reshke, Regan G. *CR #131*
Reuben, Arthur Emanuel *#619*
Reuben, Charles Gordon *#1220*
Reuben, Cyril George *#532*

Reynolds, Rodney George *#1386*
Reynolds, Samuel Henry *#173*
Richards, Bruce E. *CR #11*
Richards, George Henry *#551*
Richards, George *#692*
Richardson, David Hall *#640*
Richardson, Sidney Irving *#863*
Riddell, John Morrison *#497*
Riddell, Paul Alexander *#1236*
Ridout, William Edwin *#722*
Riley, Charles Frederick *#972*
Riley, Robert J. *#1732*
Rishchynski, Robert L. *CR #154*
Ritchie, Nelson Thomas *#246*
Rizk, Ashraf *#1915*
Roberts, Anthony Frederick *#1001*
Roberts, Donald Estyn *#1176*
Roberts, Vaughan Maurice *#232*
Roberts, Wm. Estyn Lovejoy *#625*
Robertson, Donald Mark *#1456*
Robertson, James *#214*
Robertson, John Clifford *#968*
Robertson, Ronald William *#1308*
Robinson, Daniel Bernard *#1931*
Robinson, Edward Keith *#508*
Robinson, Franklin Joseph *#318*
Robinson, Gregory G. *#1725*
Robinson, Ian David *#1472*
Robinson, Murray William *#745*
Robinson, Wallace Franklin *#952*
Robinson, William *#7*
Roccaforte, Alfonso *#1587*
Rodger, Douglas Mann *#1247*
Rody, Eric *#1804*
Rody, Leo Talson Elliot *#1096*
Roe, Bert James *#617*
Roeser, Heinrich Ludwig S. *#1140*
Roger, Alec *#426*
Roger, John *#248*
Rogers, Charles Dawson *#1144*
Rogers, Grenville Thomas *#750*
Rogers, Richard Birdsall *#161*
Rolfson, Orville *#491*
Rombough, William R. *#13*
Rooth, Carl Joseph *#1361*
Rorke, Louis Valentine *#263*
Rose, Hugh Grant *#577*
Rose, Lloyd Bradley *#664*
Ross, Burns McKay *#758*
Ross, George *#166*
Ross, Joseph Edmund *#267*
Ross, Kenneth George *#401*
Ross, Lachlan Eoin McRae *#1227*
Ross, Richard Harland *#1070*
Rouse, Tracy R. *#1856*
Routly, Herbert Thomas *#379*
Rowan, Robert Edwin *#1562*
Rowe, Duncan S. *CR #146*
Rowe, Timothy *#1362*
Roy, André Roger *#1910*
Royce, Charles Donald *#788*
Rubidge, Tom S. *#14*
Rubidge, Walter F. Brendon *#449*
Rudnicki, Les S. *#1733*
Rueb, Erich Robert *#1541*
Rueb, Guenther *#1191*
Rupert, John Glen *#1130*
Russell, Alexander Lord *#116*
Rutherford, Frank Nicol *#370*
Ruttan, Steven C. *#1671*
Ruuska, Seppo Mikael *#1416*
Ryan, Allan Louis *#1019*
Ryan, Wilfred Joseph *#724*

• S •

Sadowski, Douglas J. *CR #46*
Salb, Thomas J. *#1875*
Salna, Peter *#921*
Salna, Robert *#1523*
Salter, David F. *CR #147*
Salter, William James *#1168*
Salzer, Eric G. *#1894*
Sam-Guindon, Kathryn *#1544*
Sanders, David G. *#1685*
Sanderson, Daniel Leavens *#191*
Sani, Anthony P. *CR #12*
Sankey, Alister D. *#1842*
Sankey, Villiers *#154*
Saunders, Bryce Johnston *#204*
Sauvé, Peter Isaac Richard *#1260*
Sauvé, Sheryn I. *CR #122*
Sawayama, Gary Takayuki *#1300*
Sawhney, Jasbir Kumar *#1404*
Sawicki, Walter Raymond *#833*
Scane, Thomas *#87*
Schaeffer, Fred *#922*
Schan, Raymond Armand *#1090*
Schofield, Milton C. *#6*
Schoolkate, Louis *#1711*
Schultz, David Paul John *#1083*
Schwitzer, John Edward *#308*
Scott, John Russell *#528*
Scott, Reginald Raymond *#1028*
Seager, Edmund *#60*
Searles, David Bruce *#1225*
Sears, John Edgar *#662*
Seawright, Thomas Charles *#974*
Seawright, William Alexander *#570*
Secord, James Michael *#1418*
Seely, James Alexander *#904*
Segaran, Nath *#1978*
Seguin, Ryan William *#1920*
Seibert, Frederick V. *#447*
Selby, Henry Walter *#130*
Seleem, Nahed N. *#1890*
Senkus, Tomas A. *#1611*
Serack, William Arthur *#1482*
Setterington, Nancy Lorraine *#1234*
Setterington, William Jon *#1040*
Sevigny, Robert F. *CR #189*
Sewell, Eric Dearden *#581*
Sewell, Harry DeQuincy *#626*
Sewell, Henry C. DeQuincy *#471*
Sewell, Henry DeQuincy *#213*
Sewell, Robert George *#646*
Sexton, Christopher Anthony *#1188*
Seymour, Horace Llewellyn *#385*
Shanmugarajah, Tharmarajah *#1857*
Shantz, Murray R. *#1686*
Shaver, John Edward Hartnoll *#667*
Shaw, Charles Aeneas *#151*
Shaw, John Henry *#328*
Sheehy, Paul John *#1633*
Shelp, V. Andrew *#1719*
Sherman, Ruyter Stinson *#261*
Shier, Donald Bruce *#975*
Shipman, Harry A. Kendall *#1175*
Shipman, Jeffrey P. *#1697*
Shivkumar, Thakoordial *CR #34*
Shoemaker, Robert Leander *#716*
Shugg, Ronald Glenn *#1229*
Sibbett, William Algernon *#453*
Sibthorp, Raymond *#1904*
Siddiqi, Fuzail A. *#1524*
Silburn, James L. *CR #124*
Silvera, Roger A. Hastings *#1139*
Silvester, George Ernest *#284*
Simanovskis, Janis *#1284*
Simmonds, Douglas Andrew *#1473*
Simone, Roy *#1698*
Simpson, Alan James *#1038*
Simpson, Arthur Ferguson *#1012*
Simpson, George Albert *#85*
Simpson, James S. *CR #57*
Simpson, Michael J. *#1794*
Simpson, Moir Neil *#683*
Simpson, Walter J. *#1518*
Sinclair, John Peter *#1193*
Sing, Josiah Gershom *#162*
Singh, Tirbhowan *#1687*
Sinnis, Spiro *#1673*
Skandarajah, Nagalingam *#1717*
Skranda, Aivars *#1302*
Skuro, Peter M. *#1699*
Slater, Nicholas John *#427*
Slessor, David R. *#930*
Smeaton, George Blakbell *#1165*
Smeeton, John Boyce *#1194*
Smeets, Peter Willy Maria *#1166*
Smerdon, Glen Frederick *#1395*
Smith, Angus *#302*
Smith, Campbell Taylor *#541*
Smith, Charles Campbell *#373*
Smith, Donald Campbell *#681*
Smith, Donald G. *#1599*
Smith, Donald James *#1205*
Smith, Edwin Stuart *#1248*
Smith, Frank Harding *#806*
Smith, G. Anthony *#1600*
Smith, George Eldon *#694*
Smith, George *#176*
Smith, Harold Albert *#1011*
Smith, Harvey Morris *#1091*
Smith, Henry *#64*
Smith, Hubert C. *CR #174*
Smith, Hugh George *#1088*
Smith, Ian D. *CR #125*
Smith, J. Edward *CR #175*
Smith, James Andrew *#1448*
Smith, James Herbert *#345*
Smith, Kevin R.D. *#1960*
Smith, Laurence George *#1080*
Smith, Oliver *#500*
Smith, Peter Grenville *#1287*
Smith, Ralph A. *CR #66*
Smith, Ralph Angus *#898*
Smith, Robert Rudolph *#652*
Smith, Ronald Howard *#1296*
Smith, Steven J. *CR #103*
Smith, Sydney George *#573*
Smith, Walter *#374*
Smith, William Douglas *#1403*
Sneath, James Robert *#927*
Snell, William D. *#1601*
Snow, Ernest Arthur *#412*
Snucins, Erik P. *#1712*
Speight, Thomas Bailey *#185*
Sperling, Ernest G. *#1799*
Spragge, Godfrey Lancelot *#885*
Springate, Mark C. *CR #176*
Spry, William *#42*
Square, Berkeley Grant *#1061*
Squire, Richard Herbert *#304*
Sroka, Anthony V. *CR #182*
Srom, Jaromir *CR #52*
Stacey, Albert George *#383*
Stanciu, Christian *#1715*
Stanton, John Arthur *#1143*
Stanton, John Chester *#1570*
Stanton, William Charles *#953*
Starcevic, Dario *#1850*
Starr, William Beverly *#848*
Stassen, Bastian John *#1365*
Staszak, Tadeusz *#1036*
Statham, Sidney James *#1469*
Stauffer, Charles Edward *#656*
Stauskas, Tony *#1672*
Steele, David Taylor *#357*
Steele, Edward Charles *#255*
Steele, Gerald Francis *#1214*
Steele, Ira John *#415*
Stel, Joseph *#1312*
Stephen, Adam Michael F. *#1948*
Stephenson, Robert Terrence *#1113*
Stewart, Elihu *#112*
Stewart, George Alexander *#19*
Stewart, John L. *CR #53*
Stewart, John *#242*
Stewart, Lionel Douglas Noble *#408*
Stewart, Louis Beaufort *#189*
Stewart, Robert Craig *#1164*
Stewart, Ronald James *#1457*
Stewart, Walter Edgar *#274*
Stickings, Anthony Owen *#1071*
Stidwill, Francis Cletus *#1486*
Stidwill, Grant T. *#1769*
Stidwill, Kirk L. *#1588*
Stidwill, Leonard Putnam *#644*
Stinson, Clifford Elwood *#899*
Stinson, Robert Bruce *#737*
Stirling, Robert D. *#1513*
Stocker, Robert Allen *#1178*
Stojanovic, Svetomir *#1843*
Stones, Gerald Paul *#1330*
Stones, Michael J. *CR #27*
Strain, Robert Frederick *#747*
Strange, Henry *#1*
Strathern, John *#137*
Street, James Cunard *#454*
Stretton, William Donald *#1197*
Stringer, David B. *#1783, CR #58*
Stringer, Peter John *#1444*
Strong, Meredith C. *#1796*
Strongman, Charles Thomas *#1428*
Stubberfield, William Charles *#1431*
Stull, William Walter *#329*
Sturm, Arnold Gustav *#1397*
Suda, Philip S. *#1589*
Summers, Gordon Foster *#387*
Sun, Patrick X. *CR #127*
Sundar, Ganesh *#1969*
Sunstrum, Nelson Joseph *#900*
Suppa, Pasquale *#1659*
Sury, Leslie *#1235*
Sussman, Raphael *CR #186*
Sutcliffe, Homer Wilson *#388*
Sutherland, Bloss J. *#1858*
Sutherland, Norman Elliott *#1435*
Sutherland, Patrick J. Morrison *#1055*
Sweezey, David Edward Caird *#866*
Swift, Phillip S. *#1879*
Sylvester, John Leonard *#721*
Syme, Barrie Gordon *#1057*
Szentesi, Andras *CR #72*

• T •

Tacium, Edward Carl *#1032*
Taggart, Walter Ross *#1326*
Talbot, Jeffrey P. *#1862*

Register of Ontario Surveyors, 1892-2016 continued

Tallieu, John *#1602*
Tamblyn, Gerald A. *#905*
Tamblyn, William Bryan *#1426*
Tarantino, Giovanni *CR #185*
Tarasick, Walter Peter *#923*
Tarr, Brian Aubrey *#1141*
Tarvydas, Albinas *#937*
Tate, Henry William *#434*
Taucer, Walter *CR #193*
Taurins, Normans V. *#1734*
Taylor, Charles Gray *#643*
Taylor, Edward T. Charles *#834*
Taylor, William Emerson *#428*
Taylor, William E. Gladstone *#838*
Taylor, William Verner *#306*
Teetzel, Stewart Loren *#1223*
Tench, William Eastwood *#155*
te Riele, Hendrikus Joannes *#1371*
Terry, Michael J. *#1514*
Thachuk, Bruce E. *CR #138*
Thaler, Robert Claire *#1563*
Theyers, Ronald George *#924*
Thom, Kevin S. *#1795*
Thompson, Augustus Clifford *#57*
Thompson, Chris M. *#1700*
Thompson, Edward F. *CR #28*
Thompson, Ivan Lloyd *#769*
Thomsen, Paul R. *#1844*
Thornton, Michael John *#1412*
Thorpe, Peter *#1603*
Tieman, Andrea Ellen *#1635*
Tiernan, Joseph Martin *#216*
Tierney, Kevin M. *CR #148*
Ting, Eric T. *#1823*
Tischart, Stephen Anthony *#1618*
Todgham, Herbert Harvey *#682*
Tomaszewski, Henry *#1911*
Tomkins, Lawrence Preston *#867*
Tomlinson, Robert Douglas *#1110*
Tompsett, Bartlett *#962*
Topham, Robert James *#586*
Torbicki, Lydia M. *CR #129*
Torrance, Paul Hinton *#1340*
Townsend, David Thomas *#363*
Tracey, Thomas Henry *#110*
Traynor, Isaac *#117*
Trepanier, Joseph Bernard *#755*
Trivers, Colin George *#1279*
Trollope, Donald Herbert *#710*
Troup, Gordon Newell *#649*
Truchon, Mel *#1938*
Tsai-Laing, Cynthia *#1590*
Tulloch, David *#1954*
Tulloch, Mark Kenneth *#1905*
Tulloch, Michael F. *#1604*
Turnbull, Thomas *#159*
Turner, Howard *#1490*
Turpel, Wayne Deearr *#1348*
Tyrrell, George Carey *#579*
Tyrrell, James Williams *#207*

• U •

Uderstadt, Ernest Henry *#1131*
Unwin, Charles *#17*
Ure, Douglas Gordon *#539*
Ure, Frederick Alexander *#846*
Ure, Frederick John *#237*
Ure, Robert John *#796*
Ure, Wilfred Gordon *#518*
Urso, David Santo *#1476*

• V •

van Buskirk, William Fraser *#244*
van der Veen, Blake Campbell *#1935*
Vanderveen, Gary B. *#1777*
Van Harten, Leonard *#1303*
Van Harten, Menno Peter *#1155*
van Horssen, Frank P. *CR #81*
Van Lankveld, Ted *#1515*
van Nostrand, Arthur Jabez *#192*
van Nostrand, John *#417*
VanPelt, Herman Leendert *#890*
van Skiver, Leighton Adelbert *#631*
Van Soeren, Eric Henry *#1471*
Vaughan, Alan Lewis *#779*
Vaughan, George Brian *#1355*
Verdun, Michael D. *#1757*
Verhaegen, J-P Marie *#1203*
Verhoef, Henriette J. *#1817*
Verhoef, Nicolaas Henkrik *#1134*
Vervenne, B. Petrus *#1445*
Vicars, John Richard Odlum *#228*
Vinklers, John *#1259*
Visser, Raymond John *#1396*
Vollebekk, Danny Ronald *#1417*
Vollick, Stephen M. *#1765*
Vorsteveld, Martinus *#1186*

• W •

Waddell, Frank Melbourne *#542*
Waddell, Malcolm J. *CR #63*
Waddell, William Henry *#356*
Wadsworth, Vernon Bayley *#80*
Wadsworth, William J. Gordon *#612*
Wagner, William *#94*
Wahba, Christopher *#1929*
Wahba, Youssef *#1845*
Walczak, Jacek *#1902*
Walker, Alfred Paverley *#186*
Walker, Darren R. *#1846*
Walker, Douglas John *#928*
Walker, James Everett *#1334*
Walker, James William *#960*
Walker, John E. *CR #94*
Wall, Francis Edward *#1369*
Wallace, Alfred Ernest *#1200*
Wallace, Charles Hugh *#258*
Wallace, Ivan Bruce *#1056*
Wallace, James Nevin *#319*
Wallace, Michael J. *CR #139*
Walsh, Thomas William *#4*
Walton, Donald Frank *#748*
Walton, Robert Lowe *#794*
Wandabense, Dane *#733*
Wannack, Robert John *#1944*
Ward, Archeson Thomas *#310*
Ward, Gilbert Edward *#566*
Ward, John McLean *#1463*
Ward, Norman Everett *#545*
Ward, Stanley Ernest *#594*
Warren, Brad Keith *#1660*
Warren, James *#86*
Warren, Robert P. *CR #73*
Wassef, Attallah M. *CR #25*
Waterman, Robert George *#1204*
Watson, John McCormack *#276*
Watson, Keith W. *#1735*
Watson, Mark T. *#1504, CR #207*
Watson, McClelland Barry *#645*
Watson, Walter Irwin *#1060*
Watt, David R. *CR #152*
Weatherald, Thomas *#28*
Weatherbe, Richard Francis *#1281*
Weatherseed, John G. *CR #79*
Weaver, Glenn Arthur *#858*
Weaver, William Frederick *#568*
Webster, Brian J. *#1770*
Webster, Frederick Theodore *#554*
Webster, John Rochard *#1052*
Webster, William Gourlay *#451*
Webster, William James *#1319*
Weekes, Abel Seneca *#262*
Weekes, Melville Bell *#330*
Wegman, George Johann *#998*
Weir, David Henry *#540*
Wells, Frederick Arthur *#364*
Welmers, Arnold P. *CR #13*
Welsman, Roger Raikes *#1035*
Werrell, Adam *#1887*
West, Horace William *#754*
West, Robert Francis *#177*
Weston, John Fillimore *#727*
Weston, Sandra L. *CR #82*
Westwood, Bruce Forrest *#1119*
Whale, Harry Raymond *#982*
Wheeler, Arthur Oliver *#180*
Wheeler, Harold Cyril George *#597*
Wheelock, Charles John *#29*
Wheelock, Charles Richard *#218*
White, Dan Francis *#1438*
White, James Kenneth *#1180*
White, James Robert *#896*
White, John Edwin *#1477*
White, Walter Russell *#465*
Whitlock, Peter J. *CR #126*
Whitson, James Francis *#222*
Whittier, Albert Ronald *#564*
Whittle, Richard A. *#1674*
Wicken, Jack Edward *#713*
Wickens, John Hammond *#703*
Wicksteed, Henry King *#219*
Wiegenbröker, Robert *#1696*
Wiggins, Thomas Henry *#273*
Wight, Cecil Douglas *#571*
Wilband, Jason P. *#1877*
Wilcox, Luke G. *#1996*
Wilde, Harold Samuel *#714*
Wilde, John Absalom *#256*
Wildman, William Neil *#967*
Wilkie, Edward Thomson *#268*
Wilkins, Claude Hughes *#502*
Wilkins, Frederick William *#141*
Wilkinson, Kenneth D. *#1758*
Wilks, George Leslie *#887*
Williams, Brian Louis Gordon *#1451*
Williams, David *#81*
Williams, Edward J. *#1675*
Williams, Kenneth A. James *#827*
Williams, Peter John *#1211*
Williams, William Henry *#578*
Willson, Alfred *#100*
Wilson, Alexander Robert *#1331*
Wilson, Frank Charles *#1043*
Wilson, George Robert *#756*
Wilson, Kenneth Paul *#1427*
Wilson, Norman Douglas *#409*
Wilson, Reid John *#1484*
Wilson, Thomas Lawrence *#780*
Wilton, David A. *#1612*
Wimmelbacher, J. Herman *#1624*
Winarski, Paul Michael *#651*
Windsor, Thomas Richard *#1218*
Winter, Henry *#20*
Winters, William Selwyn *#503*
Wintle, Albert John *#1163*
Wiseman, Kenneth Murray *#783*
Woitowich, William A. *CR #151*
Wollerman, Wayne Richard *#1409*
Wong, Minos Bartholomew *#826*
Wong, Wing-Hong Stephen *#1441*
Wood, Gordon Henry *#856*
Wood, Henry O. *#26*
Wood, James Russell *#482*
Wood, John Charles *#1189*
Woodcock, Robert Victor *#1453*
Woodland, David William *#1475*
Woodroffe, Dale A. *CR #74*
Woods, Lawrence George *#1135*
Woods, Russell Alfred *#1307*
Woods, Sidney William *#600*
Woolley, James Patrick *#1645*
Worobec, Alan J. *#1613*
Wraith, Charles H. *CR #54*
Wright, Bruce M. *CR #6*
Wright, Bruce *#728*
Wright, Eugene Peter *#1384*
Wright, George Barry *#775*
Wright, Robert Hunter *#1069*
Wright, William Murray *#849*
Wylie, David J. *#1820*
Wyman, Paul Colby *#1344*
Wyse, Robert C. *CR #104*

• Y •

Yadollahi, Seyed M. *#1866*
Yalda, Bahram *#1916*
Yarnold, William Edward *#22*
Yates, Charles Robert *#533*
Yates, Donald Floyd *#964*
Yates, George Thomas *#778*
Yates, William Charles *#711*
Yeo, Michael W. *#1807*
Yeung, Albert K. *CR #140*
Young, Alexander Campbell *#450*
Young, James Gordon *#1231*
Young, John F. Gordon *#1493*
Young, John Keith *#1446*
Young, Joseph R. *#1821*
Young, R. Kent *#1377*
Young, Timothy A. *#1505*
Yuen, John Ho-Ting *#1964*

• Z •

Zabek, Stanislaw Gustaw *#1195*
Zaborski, Michael *CR #36*
Zaharieva, Yordanka N. *#1933*
Zapata, Juan Diego *#1979*
Zeng, Zhiqiang *#1925*
Zervos, George F. *#1835*
Zettel, Robert Walter *#1357*
Zeuner, Dieter *CR #106*
Ziegler, Rolf H. Heinrich *#1082*
Zillmer, Gabriella R.J. *CR #80*
Zivko, Rudolf *#1380*
Zizek, William *#1736*
Zubek, George John *#1068*
Zurek, Wojciech J. *CR #177*

Index

– H –

– I –

– J –

– K –

– L –

– M –

– S –

– T –

– U –

– V –

– W –

– Y –

– Z –